G. Blasse, B. C. Grabmaier

Luminescent Materials

With 171 Figures and 31 Tables

Springer-Verlag
Berlin Heidelberg New York
London Paris Tokyo
Hong Kong Barcelona Budapest

Prof. Dr. G. Blasse

Debye Institute
University Utrecht
Postbox 80.000
3508 TA Utrecht
The Netherlands

Prof. Dr. B. C. Grabmaier

Siemens Research Laboratories
ZFE BT MR 22
D-81730 München
Germany

also with Debye Institute
University Utrecht

ISBN 3-540-58019-0 Springer-Verlag Berlin Heidelberg New York
ISBN 0-387-58019-0 Springer-Verlag New York Berlin Heidelberg

Library of Congress Cataloging-in-Publication Data
Blasse, G. Luminescent materials / G. Blasse, B.C. Grabmaier. p. cm.
Includes bibliographical references and index.
ISBN 3-540-58019-0. -- ISBN 0-387-58019-0 (U.S.)
1. Phosphors. 2. Luminescence. I. Grabmaier, B. C., 1935- II. Title.
QC476.7.B53 1994 620.1'1295--dc20 94-20336 CIP

© Springer.Verlag Berlin Heidelberg 1994
Printed in Germany

Typesetting with TeX: Data conversion by Lewis & Leins, Berlin
SPIN: 10187460   02/3020 - 5 4 3 2 1 0 - Printed on acid-free paper

Preface

Luminescence is just as fascinating and luminescent materials (are) just as important as the number of books on these topics are rare. We have met many beginners in these fields who have asked for a book introducing them to luminescence and its applications, without knowing the appropriate answer. Some very useful books are completely out of date, like the first ones from the late 1940s by Kröger, Leverenz and Pringsheim. Also those edited by Goldberg (1966) and Riehl (1971) can no longer be recommended as up-to-date introductions.

In the last decade a few books of excellent quality have appeared, but none of these can be considered as being a general introduction. Actually, we realize that it is very difficult to produce such a text in view of the multidisciplinary character of the field. Solid state physics, molecular spectroscopy, ligand field theory, inorganic chemistry, solid state and materials chemistry all have to be blended in the correct proportion.

Some authors have tried to obtain this mixture by producing multi-authored books consisting of chapters written by the specialists. We have undertaken the difficult task of producing a book based on our knowledge and experience, but written by one hand. All the disadvantages of such an approach have become clear to us. The way in which these were solved will probably not satisfy everybody. However, if this book inspires some of the investigators just entering this field, and if it teaches him or her how to find his way in research, our main aim will have been achieved.

The book consists of three parts, although this may not be clear from the table of contents. The first part (chapter 1) is a very general introduction to luminescence and luminescent materials for those who have no knowledge of this field at all. The second part (chapters 2-5) gives an overview of the theory. After bringing the luminescent center in the excited state (chapter 2: absorption), we follow the several possibilities of returning to the ground state (chapter 3: radiative return; chapter 4: nonradiative return; chapter 5: energy transfer and migration). The approach is kept as simple as possible. For extensive and mathematical treatments the reader should consult other books.

Part three consists of five chapters in which many of the applications are discussed, viz. lighting (chapter 6), cathode-ray tubes (chapter 7), X-ray phosphors and scintillators (chapters 8 and 9), and several other applications (chapter 10). These chapters discuss the luminescent materials which have been, are or may be used in the applications concerned. Their performance is discussed in terms of the theoretical models presented in earlier chapters. In addition, the principles of the application and the preparation of the materials are dealt with briefly. Appendices on some, often not-well-understood, issues follow (nomenclature, spectral units, literature, emission spectra).

We are very grateful to Mrs. Jessica Heilbrunn (Utrecht) who patiently typed the manuscript and did not complain too much when correction after correction appeared over many months. Miss Rita Bergt (München) was of help in drawing some of the figures. Some of our colleagues put original photographs at our disposal.

This book would not have been written without discussions with and inspiration from many colleagues over a long period of time. These contacts, some oral, some via written texts, cover a much wider range than the book itself. In the preparation of this book our communication with Drs. P.W. Atkins, F. Auzel, A. Bril, C.W.E. van Eijk, G.F. Imbusch, C.K. Jørgensen, and B. Smets has been very useful.

For many years we have enjoyed our work in the field of luminescence. We hope that this book will help the reader to understand luminescence phenomena, to design new and improved luminescent materials, and to find satisfaction in doing so.

Spring 1994   G. Blasse, Utrecht
                                                   B.C. Grabmaier, München

Table of Contents

Chapter 8 X-Ray Phosphors and Scintillators (Integrating Techniques)

Chapter 9 X-Ray Phosphors and Scintillators (Counting Techniques)

Chapter 10 Other Applications

CHAPTER 1

A General Introduction to Luminescent Materials

This chapter addresses those readers for who luminescent materials are a new challenge. Of course you are familiar with luminescent materials: you meet them everyday in your laboratory and in your home. If this should come as a surprise, switch on your fluorescent lighting, relax in front of your television set, or take a look at the screen of your computer. Perhaps you would like something more specialized. Remember then your visit to the hospital for X-ray photography. Or the laser in your institute; the heart of this instrument consists of a luminescent material. However, such a high degree of specialization is not necessary. The packet of washing powder in your supermarket also contains luminescent material.

Now that we have been reminded of how full of luminescent materials our society is, the question arises "How do we define a luminescent material?" The answer is as follows: A luminescent material, also called a phosphor, is a solid which converts certain types of energy into electromagnetic radiation over and above thermal radiation. When you heat a solid to a temperature in excess of about 600°C, it emits (infra)red radiation. This is thermal radiation (and not luminescence). The electromagnetic radiation emitted by a luminescent material is usually in the visible range, but can also be in other spectral regions, such as the ultraviolet or infrared.

Luminescence can be excited by many types of energy. Photoluminescence is excited by electromagnetic (often ultraviolet) radiation, cathodoluminescence by a beam of energetic electrons, electroluminescence by an electric voltage, triboluminescence by mechanical energy (e.g. grinding), X-ray luminescence by X rays, chemiluminescence by the energy of a chemical reaction, and so on. Note that thermoluminescence does not refer to thermal excitation, but to stimulation of luminescence which was excited in a different way.

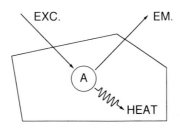

Fig. 1.1. A luminescent ion A in its host lattice. *EXC:* excitation; *EM:* emission (radiative return to the ground state); *HEAT:* nonradiative return to the ground state

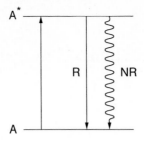

Fig. 1.2. Schematic energy level scheme of the luminescent ion A in Fig. 1.1. The *asterisk* indicates the excited state, *R* the radiative return and *NR* the nonradiative return to the ground state

In Fig. 1.1, we have drawn schematically a crystal or a grain of a photoluminescent material in order to illustrate the definition. Our system consists of a host lattice and a luminescent center, often called an activator. For example, consider the famous luminescent materials $Al_2O_3 : Cr^{3+}$ (ruby) and $Y_2O_3 : Eu^{3+}$. The host lattices are Al_2O_3 and Y_2O_3, the activators the Cr^{3+} and the Eu^{3+} ions.

The luminescence processes in such a system are as follows. The exciting radiation is absorbed by the activator, raising it to an excited state (Fig. 1.2). The excited state returns to the ground state by emission of radiation. This suggests that every ion and every material shows luminescence. This is not the case. The reason for this is that the radiative emission process has a competitor, viz. the nonradiative return to the ground state. In that process the energy of the excited state is used to excite the vibrations of the host lattice, i.e. to heat the host lattice. In order to create efficient luminescent materials it is necessary to suppress this nonradiative process.

The obvious characteristics to be measured on this system are the spectral energy distribution of the emission (the emission spectrum), and of the excitation (the excitation spectrum; which in this simple case is often equal to the absorption spectrum), and the ratio of the radiative and the nonradiative rates of return to the ground state. The latter determines the conversion efficiency of our luminescent material.

Let us for a second return to ruby ($Al_2O_3 : Cr^{3+}$). This is a beautiful red gemstone which shows a deepred luminescence under excitation with ultraviolet or visible radiation. Its spectroscopic properties were studied as early as 1867 by the famous scientist Becquerel, who used sunlight as the excitation source. He claimed that the color as well as the luminescence were intrinsic properties of the host lattice. This time, however, Becquerel was wrong. It is the Cr^{3+} ion which is responsible for the optical absorption of ruby in the visible and ultraviolet spectral regions. The host lattice Al_2O_3 does not participate at all in the optical processes. In fact Al_2O_3 is colorless. In the case of the ruby, the activator A in Fig. 1.1 is the Cr^{3+} ion and the host lattice is Al_2O_3. The host lattice of this luminescent material has no other function than to hold the Cr^{3+} ion tightly.

In many luminescent materials the situation is more complicated than depicted in Fig. 1.1, because the exciting radiation is not absorbed by the activator, but elsewhere. For example, we can add another ion to the host lattice. This ion may absorb

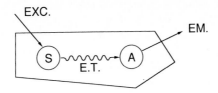

Fig. 1.3. Energy transfer from a sensitiser S to an activator A. Energy transfer is indicated by *E.T.* For further notation, see Fig. 1.1

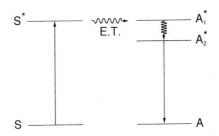

Fig. 1.4. Energy transfer from S to A. The S → S* transition is the absorption (or excitation), the A_2^* → A transition the emission. The level A_1^*, populated by energy transfer (*E.T.*), decays nonradiatively to the slightly lower A_2^* level. This prevents back transfer

the exciting radiation and subsequently transfer it to the activator. In this case the absorbing ion is called a sensitizer (see Fig. 1.3).

Another well-known example, viz. the lamp phosphor $Ca_5(PO_4)_3F : Sb^{3+}, Mn^{2+}$. Ultraviolet radiation is not absorbed by Mn^{2+}, but only by Sb^{3+}. Under ultraviolet irradiation, the emission consists partly of blue Sb^{3+} emission, and partly of yellow Mn^{2+} emission. Since the Mn^{2+} ion was not excited directly, the excitation energy was transferred from Sb^{3+} to Mn^{2+} (see Fig. 1.4). The luminescence processes can be written as follows, where $h\nu$ indicates radiation with frequency ν and the asterisk an excited state:

$$Sb^{3+} + h\nu \rightarrow (Sb^{3+})^*$$

$$(Sb^{3+})^* + Mn^{2+} \rightarrow Sb^{3+} + (Mn^{2+})^*$$

$$(Mn^{2+})^* \rightarrow Mn^{2+} + h\nu.$$

These "equations" indicate absorption, energy transfer, and emission, respectively. If the Sb^{3+} ion has no Mn^{2+} ions in its vicinity, it gives its own blue emission. For those of you who are not familiar with solids, please realize that in general the concentrations of the luminescent centers are of the order of a 1 mol.%, and that the centers are, in first approximation, distributed at random over the host lattice.

Sometimes, however, the activator concentration can be 100%. This illustrates the rather complicated nature of luminescent materials. Actually these high activator concentrations which occur in some cases were not understood for a long time, preventing progress in our understanding of luminescent materials. A famous example of such a high-concentration material is $CaWO_4$ where the tungstate group is the luminescent center. Simultaneously it is a building unit of the host lattice which consists of Ca^{2+} and WO_4^{2-} ions. This material was used for 75 years in X-ray photography, and in tungsten mines its luminescence is used to find $CaWO_4$. The miners use ultraviolet lamps to find the tungstate-rich ores by their visible luminescence! Chapter 5 discusses why a high concentration of activators is sometimes fatal for luminescence, whereas in other cases such high concentrations yield very high luminescence outputs.

In stead of exciting a low concentration of sensitizers or activators, we can also excite the host lattice. This is, for example, what happens if we excite with X rays or electron beams. In many cases the host lattice transfers its excitation energy to the activator, so that the host lattice acts as the sensitizer. Again a few examples. In $YVO_4 : Eu^{3+}$ ultraviolet radiation excites the vanadate groups, i.e. the host lattice. The emission, however, consists of Eu^{3+} emission. This shows that the host lattice is able to transfer its excitation energy to the Eu^{3+} ions. Another example is $ZnS : Ag^+$, the blue-emitting cathode-ray phosphor used in television tubes. Ultraviolet radiation, electron beams and X rays excite the sulfide host lattice which transfers this excitation energy rapidly to the activators (the Ag^+ ions).

In spite of the fact that we did not discuss any fundamental background (this will be done in Chapters 2-5), you have met by now the more important physical processes which play a role in a luminescent material:

- absorption (excitation) which may take place in the activator itself, in another ion (the sensitizer), or in the host lattice (Chapter 2)
- emission from the activator (Chapter 3)
- nonradiative return to the ground state, a process which reduces the luminescence efficiency of the material (Chapter 4)
- energy transfer between luminescent centers (Chapter 5).

After this short, general introduction into the operation of a luminescent material, we now turn to a similar introduction to the applications of luminescent materials.

Photoluminescence is used in fluorescent lamps. This application was even used before the Second World War. The lamp consists of a glass tube in which a low-pressure mercury discharge generates ultraviolet radiation (85% of this radiation consists of 254 nm radiation). The lamp phosphor (or a mixture of lamp phosphors) is applied to the inner side of the tube. This phosphor converts the ultraviolet radiation into white light. The efficiency of conversion of electricity to light is in a fluorescent lamp considerably higher than in an incandescent lamp.

The introduction of rare-earth activated phosphors in fluorescent lamps during the last decade has improved the light output and the colour rendering drastically. As a consequence this type of lighting is no longer restricted to shops and offices, but is now also suitable for living rooms. It is interesting to note that in this way chemical elements which for a long time have been considered as rare, peculiar, and hard to separate, have penetrated our houses. A modern fluorescent lamp contains the following rare earth ions: divalent europium, trivalent cerium, gadolinium, terbium, yttrium and europium. You will find more about this important application of photoluminescence in Chapter 6.

One can hardly imagine life today without cathode-ray tubes. Think of your television set or your computer screen. In a cathode-ray tube the luminescent material is applied on the inner side of the glass tube and bombarded with fast electrons from the electron gun in the rear end of the tube. When the electron hits the luminescent material, it emits visible light. In the case of a colour television tube there are three electron guns, one irradiating a blue-emitting luminescent material, so that it creates a blue pictures, whereas two others create in a similar way a green and a red picture.

One fast electron creates in the luminescent material many electron-hole pairs which recombine on the luminescent center. This multiplication is one of the factors which have determined the success of the cathode-ray tube as a display. It will be clear that the luminescent materials applied belong to the class of materials where excitation occurs in the host lattice. They will be discussed in Chapter 7 where we will also deal with materials for projection television. In this way the display screen can have a diameter of 2 m. This application puts requirements on luminescent materials which are hard to satisfy.

Let us now turn to materials which are able to convert X-ray irradiation into visible light. Röntgen discovered X rays in 1895, and realized almost immediately that this type of radiation is not very efficient at exposing photographic film, because the film does not absorb the X rays effectively. As a consequence long irradiation times are required. Nowadays we know that this is bad for the patient. There is also a practical objection against long irradiation times: the patient is a moving object (he breathes and may, in addition, make other movements), so that sharp pictures can only be obtained if the irradiation time is short.

Therefore Röntgen initiated a search for luminescent materials which absorb X rays efficiently and convert their energy into radiation which is able to blacken the photographic film. Soon it was found that $CaWO_4$ with a density of 6.06 g.cm^{-3} was able to do so. This compound was used for a very long time in the so-called X-ray intensifying screens. A schematic picture of X-ray photography with this method is given in Fig. 1.5.

Just as in the field of lamp phosphors and (partly) cathode-ray phosphors, $CaWO_4$ lost its leading position to rare-earth activated X-ray phosphors (see chapter 8). As a salute to this old champion, but also for your information, we give in Fig. 1.6 the crystal structure of $CaWO_4$ which illustrates the build-up of the lattice from Ca^{2+} ions and luminescent WO_4^{2-} groups, and in Fig. 1.7. an electron micrograph of a commercial $CaWO_4$ powder.

A recent development in this field is the introduction of storage phosphors. These materials have a "memory" for the amount of X rays which has been absorbed at a given spot of the screen. By scanning the irradiated screen with an (infra)red laser, visible luminescence is stimulated. Its intensity is proportional to the amount of X rays absorbed. Chapter 8 discusses how to produce such materials and the physics behind these phenomena.

As an example of a more specialized character, we should mention the case of X-ray computed tomography. This method of medical radiology generates cross-sectional images of the interior of the human body (see Fig. 1.8). Besides the X-ray source, the key component is the detector consisting of about 1000 pieces of a luminescent solid (crystals or ceramics) which are connected to photodiodes or -multipliers. The

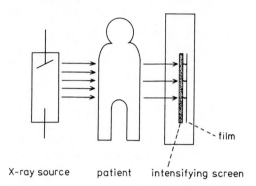

X-ray source   patient   intensifying screen

film

Fig. 1.5. Schematic representation of a medical radiography system based on the use of an intensifying screen

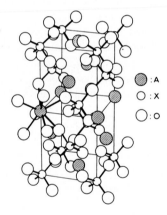

: A
: X
: O

Fig. 1.6. Crystal structure of $CaWO_4$ (scheelite). The general formula is AXO_4 where A are the larger and X the smaller metal ions

luminescent material has to show sharply defined properties in order to be acceptable for this application.

After these examples of X-ray photography, you will not be surprised to learn that α and γ radiation can also be detected by luminescent materials, which, in this case, are usually called scintillators and are often in the form of large single crystals. The applications range from medical diagnostics (for example positron emission tomog-

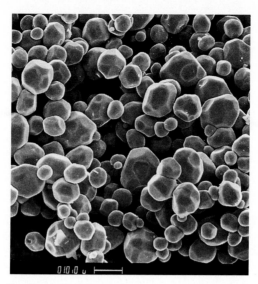

Fig. 1.7. $CaWO_4$ powder phosphor (1000×)

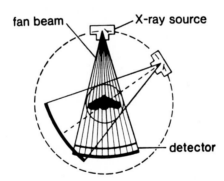

Fig. 1.8. The principle of X-ray computed tomography. The patient is in the center of the picture. Source and detector rotate around the patient

raphy (PET)) to nuclear and high-energy physics. A spectacular application in the latter field is the use of 12 000 crystals of $Bi_4Ge_3O_{12}$ ($3 \times 3 \times 24$ cm^3) as a detector for electrons and photons in the LEP machine at CERN (Geneva). Scintillators will be discussed in Chapter 9.

As a matter of fact there are many other types of application. Some of these are discussed in Chapter 10. Before you get the impression that (the application of)

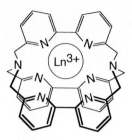

Fig. 1.9. The [Ln $\subset$ bpy.bpy.bpy]$^{3+}$ cryptate

luminescence is restricted to solids, we should pick out one of these, viz. fluoro-immunoassay, which is based on luminescent molecules. This is a method used in immunology in order to detect biomolecules. The method is superior to other methods (such as those using radioactive molecules) as far as sensitivity and specificity are concerned. It is used particularly in the clinical investigation of compounds in low concentration and consists of the labelling of samples with luminescent species and the measurement of their luminescence.

One of the molecules which plays a role in this field is depicted in Fig. 1.9. The luminescent species is the Eu^{3+} ion which we met already above. It is surrounded by a cage containing molecules of bipyridine. The whole complex is called a cryptate and its formula is written as [Eu $\subset$ bpy.bpy.bpy]$^{3+}$. The cage protects the Eu^{3+} ion against the (aqueous) surroundings which tries to quench the luminescence. If this cryptate is excited with ultraviolet radiation, the bipyridine molecules absorb the exciting radiation and transfer their excitation energy subsequently to the Eu^{3+} ion which then shows its red luminescence.

Coordination chemists call this transfer from bipyridine to Eu^{3+} an antenna effect. It is of course exactly the same phenomenon which we described above (Fig. 1.3). In solid state research the effect is simply called energy transfer. In Chapter 10 we will see that the physics of molecules like cryptates is very similar to that of solids like those depicted in Fig. 1.3.

Finally that intriguing application yielding the laser. In luminescence the radiative decay of the excited state to the ground state occurs by spontaneous emission, i.e. the emission processes on different activator ions are not correlated. If, by some means, the majority of the luminescent ions are in the excited state (this situation is called population inversion), a single spontaneously emitted photon (quantum of radiation) may stimulate other excited ions to emit. This process is called stimulated emission. It is monochromatic, coherent and non-divergent. Laser action depends on emission by a stimulated process. Actually the word laser is an acronym for light amplification by stimulated emission of radiation. This book does not deal with lasers or laser physics. However, we will deal with the material where the stimulated emission is generated if this is useful for our purpose. Every laser material is after all also a luminescent material.

If you are surprised, remember the gemstone ruby ($Al_2O_3 : Cr^{3+}$) mentioned in the beginning of this chapter. Becquerel started long ago to study its luminescence and spectroscopy. Ruby was and is the start of several interesting phenomena in solid state physics. Probably the most important of these is the first solid state laser which was based on ruby (Maiman, 1960). This illustrates the connection between luminescence and lasers. On the other hand, in more recent years it has only been possible to unravel and understand luminescence processes by using laser spectroscopy.

This introductory chapter should have stimulated or excited you to read on and to become acquainted with the physics and the chemistry of luminescence and luminescent materials.

References

Chapter 1 does not have specific literature references in view of its general and introductory character. Those who are invited by this chapter to improve their knowledge of the basic physics and chemistry of energy levels, ions, spectroscopy and/or solid state chemistry, are referred to general textbooks in the field of physical and inorganic chemistry. Our personal preferences are the following: P.W. Atkins (1990) Physical Chemistry, 4th edn, Oxford University Press; and D.F. Shriver, P.W. Atkins and C.H. Langford (1990) Inorganic Chemistry, Oxford University Press (chapters 14 and 18). The ruby history has been reviewed by G.F. Imbusch (1988) In: W.M. Yen and M.D. Levenson (eds) Lasers, Spectroscopy and New Ideas. Springer Berlin Heidelberg New York. Those who want to judge the great progress in luminescent materials during the last few decades should compare this book with the review paper by J.L. Ouweltjes, Luminescence and Phosphors, Modern Materials 5 (1965) p. 161.

CHAPTER 2

How Does a Luminescent Material Absorb Its Excitation Energy?

2.1 General Considerations

A luminescent material will only emit radiation when the excitation energy is absorbed. These absorption processes will be the subject of this chapter with stress on excitation with ultraviolet radiation. The emission process will be treated in the next chapter.

Let us consider an optical absorption spectrum of a well-known luminescent material, viz. $Y_2O_3 : Eu^{3+}$. Figure 2.1 presents this spectrum in a schematic form. Starting on the long wavelength (i.e. low energy) side, the following features can be noted:

- very weak narrow lines
- a broad band with a maximum at 250 nm
- a very strong absorption region for $\lambda \le 230$ nm.

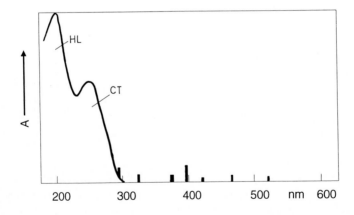

Fig. 2.1. Schematic representation of the absorption spectrum of $Y_2O_3 : Eu^{3+}$. The value of A gives the absorption strength in arbitrary units. The narrow lines are transitions within the $4f^6$ configuration of Eu^{3+}, CT is the Eu^{3+}-O^{2-} charge-transfer transition, and HL the host lattice (Y_2O_3) absorption

The absorption spectrum of pure Y_2O_3 shows only the latter absorption. Therefore the lines and the 250 nm band must be due to Eu^{3+}, and the absorption at $\lambda < 230$ nm to the host lattice Y_2O_3.

Let us now consider the excitation spectrum of the Eu^{3+} emission of Y_2O_3 : Eu^{3+}. Such a spectrum yields the luminescence output as a function of the exciting wavelength, so that there should be a correlation with the absorption spectrum. The excitation spectrum of Y_2O_3 : Eu^{3+} shows a striking agreement with the absorption spectrum of Y_2O_3 : Eu^{3+}. This means the following:

- if the Eu^{3+} ions are excited directly (sharp lines, 250 nm band), luminescence from Eu^{3+} is observed. Less trivial is the second observation:
- if the host lattice Y_2O_3 is excited ($\lambda \leq 230$ nm), luminescence from Eu^{3+} is observed. Consequently, the excitation energy absorbed by the host lattice Y_2O_3 is transferred to the activator Eu^{3+}. These transfer processes will be considered later in detail (Chapter 5). For the moment it is important to realize that absorption of excitation energy is not restricted to the activator itself, but can also occur elsewhere.

High-energy excitation always excites the host lattice. Examples are fast electrons, γ rays, X rays. Direct excitation of the activator is only possible with ultraviolet and/or visible radiation. The phosphor Y_2O_3 : Eu^{3+}, for example, is excited in the activator itself (the 250 nm band) when applied in a luminescent lamp (254 nm excitation), but in the host lattice when applied as a cathode-ray or X-ray phosphor.

In this chapter we are dealing mainly with absorption of ultraviolet or visible radiation, because its wavelength can be easily varied and this indicates exactly what and where we are exciting. As H.A. Klasens, a pioneer in luminescent materials, used to say: ultraviolet excitation compares to striking one key of the piano, cathode-ray or X-ray excitation compares to throwing the piano down the stairs.

Let us now return to the absorption spectrum of Figure 2.1. A central problem of absorption spectra can immediately be noted, viz. why are certain spectral features so narrow and others so broad, and why have some such a low and others such a high intensity. To these questions we will give an answer in a simplified form. Detailed presentations can be found in many books [1-3].

The shape of an optical absorption band, i.e. narrow or broad, can be explained using the configurational coordinate diagram. Such a diagram shows the potential energy curves of the absorbing center as a function of a configurational coordinate. This coordinate describes one of the vibrational modes of the centre involved. We will follow the method of many textbooks in considering the vibrational mode in which the central metal ion is at rest and the surrounding ligands are moving in phase away from the metal ion and coming back. This is the so-called symmetrical stretching (or breathing) mode. Figure 2.2 gives a schematic example.

The configurational coordinate diagram for this mode reduces to a plot of the energy E versus the metal-ligand distance R, since R is the structural parameter which varies during the vibration. In doing so, the reader should realize that we neglect all other vibrational modes. In older textbooks this was considered to be a justified approximation, but later it was shown to be a dangerous one, because it neglects distortions in the excited state. These often occur, as we will see later in this book.

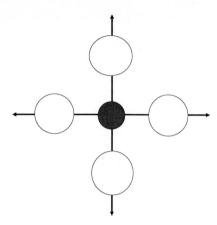

Fig. 2.2. Symmetrical stretching vibration of a square-planar complex. The ligands (open circles) move in phase from and to the central metal ion

Figure 2.3. shows a configurational coordinate diagram where E is plotted versus R. Consider first the curve for the lowest state, the ground state. Its shape is parabolic with a minimum at R_0. This parabolic shape follows from the fact that the vibrational motion is assumed to be harmonic, i.e. the restoring force F is proportional to the displacement: $F = -k(R - R_0)$. A force of this form corresponds to a potential energy whose dependence on R is parabolic: $E = \frac{1}{2}k(R - R_0)^2$. The minimum R_0 of the parabola corresponds to the equilibrium distance in the ground state.

The quantum mechanical solution of this problem (known as the harmonic oscillator) yields for the energy levels of the oscillator $E_v = (v + \frac{1}{2})h\nu$, where v = 0, 1, 2, ... and ν is the frequency of the oscillator. Part of these levels have been drawn in Figure 2.3. For a simple derivation, see e.g. Ref. [4].

The wave functions of these vibrational levels are also known. For our purpose the more important information is that in the lowest vibrational level (v = 0) the highest probability of finding the system is at R_0, whereas for high values of v it is at the turning points, i.e at the edges of the parabola (like in the classic pendulum) (see Figure 2.4.).

What has been said about the ground state holds also for the excited states: in the E-R diagram they occur also as parabolas, but with different values of the equilibrium distance (R_0') and force constant (k'). These differences are due to the fact that in the excited state the chemical bond is different from that in the ground state (often weaker). This is also shown in Figure 2.3, where the parabolas are shifted relative to each other over a value ΔR.

Before considering optical absorption processes in the configurational coordinate model, we would like to draw attention to the fact that what we have done is consider the interaction between the absorbing metal ion and the vibrations of its surroundings. Transitions between two parabolas are electronic transitions. Our model enables us, in principle, to consider the interaction between the electrons and the vibrations of the

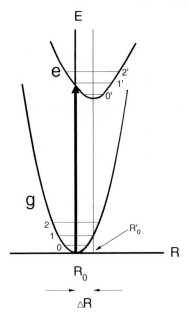

Fig. 2.3. Configurational coordinate diagram (see also text). The ground state (g) has the equilibrium distance R_0; the vibrational states $v = 0, 1, 2$ are shown. The excited state (e) has the equilibrium distance R_0'; the vibrational states $v' = 0, 1, 2$ are shown. The parabola offset is $\Delta R (= R_0' - R_0)$

optical centre under consideration. Actually the value of $\Delta R = R_0' - R_0$ is a qualitative measure of this interaction.

In optical absorption the centre is promoted from its ground state to an excited state. How is such a transition described in the configurational coordinate diagram of Figure 2.3? It is important to realize that optical transitions occur in this diagram as vertical transitions. The reason for this is that a transition from the ground state to the excited state is electronic, whereas horizontal displacements in this diagram are nuclear, the distance R being an internuclear distance. Since the electrons move much faster than the nuclei, the electronic transition takes, in good approximation, place in static surroundings. This implies a vertical transition in Figure 2.3. The nuclei take their appropriate positions only later (next chapter).

The optical absorption transition starts from the lowest vibrational level, i.e. $v = 0$. Therefore, the most probable transition occurs at R_0 where the vibrational wave function has its maximum value (see Figure 2.5). The transition will end on the edge of the excited state parabola, since it is there that the vibrational levels of the excited state have their highest amplitude. This transition corresponds to the maximum of the absorption band. It is also possible, although less probable, to start at R values larger or smaller than R_0. This leads to the width of the absorption band (Figure 2.5), because for $R > R_0$ the energy difference of the transition will be less than for $R = R_0$, and for $R < R_0$ it will be larger.

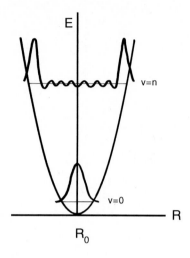

Fig. 2.4. The vibrational wave functions for the lowest vibrational level (v = 0) and a high vibrational level (v = n)

If $\Delta R = 0$, the two parabolas lie exactly above each other and the band width of the optical transition vanishes: the absorption band becomes a narrow line.

It can be simply shown [1,4] that the probability for an optical transition between the v = 0 vibrational level of the ground state and the v = v' vibrational level of the excited state is proportional to

$$\langle e|r|g\rangle\langle\chi_{v'}|\chi_0\rangle. \qquad (2.1)$$

Here the functions e and g present the electronic wave functions of the excited state and the ground state, respectively; r is the electric-dipole operator driving the transition; and χ are the vibrational wave functions. In order to consider the whole absorption band, one has to sum over v'.

The first part of eq. (2.1) is the electronic matrix element, which is independent of the vibrations. The second part gives the vibrational overlap. The former gives the intensity of the transition as will be shown below, the latter determines the shape of the absorption band.

The latter statement can be illustrated as follows. When $\Delta R = 0$, the vibrational overlap will be maximal for the levels v = 0 and v' = 0, since the vibrational wave functions involved have their maxima at the same value of R, viz. R_0. The absorption spectrum consists of one line, corresponding to the transition from v = 0 to v' = 0. This transition is called the zero-vibrational or no-phonon transition, since no vibrations are involved. If, however, $\Delta R \neq 0$, the v = 0 level will have the

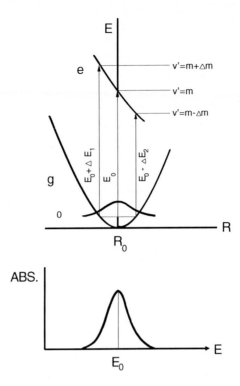

Fig. 2.5. The optical absorption transition between two parabolas which have an offset relative to each other in the configurational coordinate diagram consists of a broad absorption band. See also text

maximal vibrational overlap with several levels $v' > 0$, and a broad absorption band is observed.

The broader the absorption band, the larger the value of ΔR. The width of an absorption band informs us immediately how large the difference in ΔR (and how large the difference in chemical bonding) between the excited state and the ground state is.

It is usual to call the $\Delta R = 0$ situation the weak-coupling scheme, $\Delta R > 0$ the intermediate-coupling scheme, and $\Delta R \gg 0$ the strong-coupling scheme. The word coupling relates to the coupling between the electrons and the vibrations of the center considered. The value of ΔR measures the strength of this interaction.

At higher temperature, the initial state may also be one with $v > 0$. This results in band broadening. This well-known phenomenon is easily explained by the configurational coordinate diagram.

Now we return to the intensity of the optical absorption transition which is contained in the matrix element $\langle e|r|g \rangle$. Not every possible transition between g and e does occur as an optical transition, since these are gouverned by selection rules.

Here we should mention two important selection rules, viz.

- the spin selection rule which forbids electronic transitions between levels with different spin states ($\Delta S \neq 0$)
- the parity selection rule which forbids electronic (electric-dipole) transitions between levels with the same parity; examples are electronic transitions within the d shell, within the f shell, and between the d and the s shells.

In solids, the selection rules can seldom be considered as absolute rules. The situation is reminiscent of towns with low traffic morals: when the traffic light is green everyone crosses (allowed transition, no selection rules), at red there are still a few people who cross against the rules (forbidden transition, but selection rule slightly relaxed). The relaxation of selection rules is connected to wavefunction admixtures into the original, unperturbed wave functions. This can be due to several physical phenomena, like spin-orbit coupling, electron-vibration coupling, uneven crystal-field terms, etc. Their treatment lies outside the scope of this book. The reader is referred to Refs. [1] and [3].

Let us now return to Figure 2.1. The host lattice absorption band of Y_2O_3 is very broad and intense. This means that the excited state is strongly different from the ground state. Actually the highest occupied levels of the ground state are the $2p$ orbitals of oxygen; the lowest unoccupied levels of the excited state are a mixture of $3s$ orbitals of oxygen and $4d$ of yttrium. Without going into detail, the reader will understand that the lowest optical transition in Y_2O_3 results in large changes in chemical bonding and ΔR.

The Eu^{3+} absorption features have lower intensity. This is in the first place due to its lower concentration (there is about 1% Eu in the sample for which the spectrum of Figure 2.1 is measured). The 250 nm absorption band is a charge-transfer transition in the $Eu^{3+}-O^{2-}$ bond: an electron jumps from oxygen to europium. Consequently ΔR is large and the absorption band broad. There is no selection rule which reduces the intensity. The narrow weak lines are due to electronic transitions within the nonbonding $4f^6$ shell of Eu^{3+}. As a consequence $\Delta R = 0$, yielding narrow lines. The parity-selection rule forbids these transitions, and they are very weak indeed (often more than 10^6 times weaker than allowed transitions).

This paragraph dealt with the properties of absorbtion bands in the ultraviolet and visible region in general. Section 2.3 deals with specific cases.

2.2 The Influence of the Host Lattice

If we consider a given luminescent centre in different host lattices, the optical properties of this center are usually also different. This is not so surprising, since we change the direct surroundings of the luminescent center. In this paragraph some of these

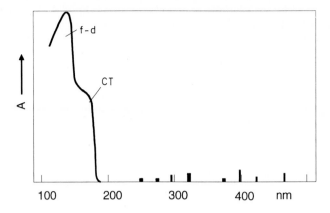

Fig. 2.6. Schematic representation of the absorption spectrum of YF_3 : Eu^{3+}. The indications CT and $f \rightarrow d$ relate to the $Eu^{3+}-F^-$ charge-transfer and the $Eu^{3+}4f^6 \rightarrow 4f^55d$ transitions, respectively

effects will be considered. They are of prime importance in the materials science of luminescent materials: if we would understand how the luminescence properties of an optical center depend on the host lattice, it would be easy to predict all the luminescent materials.

To illustrate the influence of the host lattice on the optical absorption of a center, we consider the absorption spectrum of YF_3 : Eu^{3+} (Fig. 2.6), and compare it with that of Y_2O_3 : Eu^{3+} (Fig. 2.1). The following differences are immediately obvious.

- the host lattice absorption band of Y_2O_3 has disappeared. That of YF_3 is not observed; it is situated at even shorter wavelengths than given in Figure 2.6.
- the charge-transfer absorption band of Y_2O_3 : Eu^{3+} has disappeared. In YF_3 : Eu^{3+} it appears at about 150 nm. This shows that it takes much more energy to remove an electron from an F- ion than from an O^{2-} ion.
- in addition the absorption spectrum of YF_3 : Eu^{3+} shows the allowed $4f \rightarrow 5d$ transitions of Eu^{3+} at about 140 nm.
- in both absorption spectra the transitions within the $4f^6$ configuration of Eu^{3+} are observed as sharp and weak lines. Since the $4f$ electrons are well shielded from the surroundings by completely filled $5s$ and $5p$ orbitals, the influence of the surroundings of the Eu^{3+} ion is expected to be small indeed.

Careful inspection of these narrow absorption lines shows clear differences which cannot be seen in the rough Figs. 2.1 and 2.6. In the fluoride they are even less intense than in the oxide, in the fluoride their spectral positions tend to be at slightly higher energy than in the oxide, and their splitting pattern under high resolving power is different in the two compositions.

Let us now consider the main factors responsible for different spectral properties of a given ion in different host lattices. The first factor to be mentioned is covalency [5-8]. For increasing covalency the interaction between the electrons is reduced, since they spread out over wider orbitals. Consequently, electronic transitions between energy

Table 2.1. The nephelauxetic effect for Bi^{3+} and Gd^{3+}. Covalency increases from top to bottom.

Bi^{3+} [9]		Gd^{3+} [10]	
Host lattice	$^1S_0 - ^3P_1$ $(cm^{-1})^*$	Host Lattice	$^8S \rightarrow ^6P_{7/2}$ $(cm^{-1})^{**}$
YPO_4	43.000	LaF_3	32.196
YBO_3	38.500	$LaCl_3$	32.120
$ScBO_3$	35.100	$LaBr_3$	32.096
La_2O_3	32.500	$Gd_3Ga_5O_{12}$	31.925
Y_2O_3	30.100	$GdAlO_3$	31.923

* lowest component of the $6s^2 \rightarrow 6s6p$ transition
** lowest transitions within the $4f^7$ shell

Table 2.2. The maximum of the Eu^{3+} charge-transfer transition in several host lattices [9].

Host lattice	Maximum Eu^{3+} CT (10^3 cm^{-1})
YPO_4	45
YOF	43
Y_2O_3	41.7
$LaPO_4$	37
La_2O_3	33.7
LaOCl	33.3
Y_2O_2S	30

levels with an energy difference which is determined by electron interaction shift to lower energy for increasing covalency. This is known as the nephelauxetic effect[1]. Table 2.1 gives two examples of a different nature, viz. for Bi^{3+} $(6s^2)$ and for Gd^{3+} $(4f^7)$. The former has a $6s^2 \rightarrow 6s6p$ absorption transition. The $6s$ and $6p$ electrons reside on the surface of the ion and the nephelauxetic effect is large. The latter has a weak absorption due to a transition within the $4f^7$ shell. This lies inside the ion, and the nephelauxetic effect is very small. In the same way the slightly higher positions of the transitions within the $4f^6$ shell of the Eu^{3+} ion in YF_3 compared to Y_2O_3 (see above) are also an illustration of this nephelauxetic effect.

Higher covalency means also that the electronegativity difference between the constituting ions becomes less, so that charge-transfer transitions between these ions shift to lower energy. This was mentioned already above, where the charge-transfer absorption band of Eu^{3+} in the fluoride YF_3 was observed to be at higher energy than in the more covalent oxide Y_2O_3. Table 2.2 gives some more examples. In sulfides europium is usually divalent [11], since the charge-transfer state of Eu^{3+} in sulfides lies at such low energy that the trivalent state is no longer stable.

Similar observations have been made for other ions with charge transfer transitions [5-7].

---

1 The word nephelauxetic means (electron)cloud expanding.

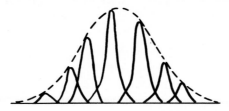

Fig. 2.7. Inhomogeneous broadening. The individual absorption transitions vary slightly from site to site in the host lattice. The broken line indicates the experimentally observed absorption spectrum

Another factor responsible for the influence of the host lattice on the optical properties of a given ion is the crystal field. This is the electric field at the site of the ion under consideration due to the surroundings. The spectral position of certain optical transitions is determined by the strength of the crystal field, the transition metal ions being the most well-known and clear example. For example, why is Cr_2O_3 green, but $Al_2O_3 : Cr^{3+}$ red, whereas both compositions have the same crystal structure? The qualitative answer is simple: in ruby ($Al_2O_3 : Cr^{3+}$) the Cr^{3+} ions (which are responsible for the color) occupy the smaller Al^{3+} sites, so that they feel a stronger crystal field than in Cr_2O_3. Therefore the optical transitions in ruby are at higher energy than in Cr_2O_3, so that the color of the two compositions is different.

In addition the crystal field is responsible for the splitting of certain optical transitions. Obvious is the following statement: different host lattices → different crystal fields → different splittings. In this way the optical center can serve as a probe of the surroundings: observed splittings yield the symmetry of the site.

Crystal fields of uneven (ungerade) symmetry are able to lift the parity selection rule. The reader should not underestimate the importance of such a, seemingly-academic, statement: there would be no color television, and no energy-saving luminescent lamps without uneven-symmetry crystal fields, as we will see later.

Up till now, it has been tacitly assumed that the surroundings and the symmetry of each center in the solid are the same. This is the case for Eu^{3+} in YF_3, the crystallographic sites of all Y^{3+} ions being equal. It should be realized that, in powders, the external, and also the internal surface may be large, and that Eu^{3+} ions near this surface experience a covalency and a crystal field which differs from the bulk. These Eu^{3+} ions have their optical transitions at energies which are slightly different from those in the bulk. Therefore the features in the spectra broaden. This is called inhomogeneous broadening (see Fig. 2.7). Point defects in the crystal structure yield also a contribution to this broadening.

The simple-looking $Y_2O_3 : Eu^{3+}$ is already more complicated, since the crystal structure offers for Y^{3+} two crystallographic sites with different symmetry. The Eu^{3+} ions are found on both sites with different spectral properties.

Pronounced inhomogeneous broadening occurs in glasses where all optical centers differ from site to site due to the absence of translational symmetry. Absorption bands in glasses are, therefore, always broader than in crystalline solids.

2.3 The Energy Level Diagrams of Individual Ions

In this section some specific (groups of) ions will be discussed more in detail as far as necessary for an understanding of the luminescence of luminescent materials. For more details the reader is referred to other books (for example, Refs [1,7,8]). For simplicity the energy level diagrams will be presented for one distance, viz. the equilibrium distance of the ground state. Therefore they contain less information than a configurational coordinate diagram (like those in Figs 2.3 and 2.5).

2.3.1 The Transition Metal Ions (d^n).

Transition metal ions have an incompletely filled d shell, i.e. their electron configuration is $d^n (0 < n < 10)$. The energy levels originating from such a configuration have been calculated by Tanabe and Sugano taking the mutual interaction between the d electrons as well as the crystal field into account. Figures 2.8–2.10 give, as an example, the results for the configurations d^1, d^3 and d^5.

On the utmost left-hand side (crystal field $\Delta = 0$) we find the energy levels of the free ion. Many of these levels split into two or more levels for $\Delta \neq 0$, as for example in a solid. The lowest level, i.e. the ground state, coincides with the x-axis. For the free ion the levels are marked ^{2S+1}L, where S presents the total spin quantum number, and L the total orbital angular momentum. Values of L may be 0 (indicated by S), 1 (P), 2 (D), 3 (F), 4 (G), etc. The degeneracy of these levels is 2L+1 and may be lifted by the crystal field. Crystal-field levels are marked ^{2S+1}X, where X may be A (no degeneracy), E (twofold degeneracy) and T (threefold degeneracy). Subscripts indicate certain symmetry properties. For more details the reader is referred to Refs [12] and [13]. The indicated nomenclature can be checked in Figs 2.8–2.10.

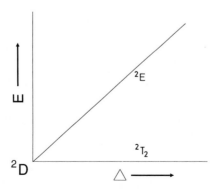

Fig. 2.8. The energy levels of the d^1 configuration as a function of the octahedral crystal field Δ. The free ion level is ^2D

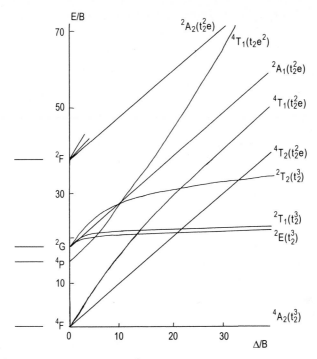

Fig. 2.9. Tanabe-Sugano diagram for the d^3 configuration. Free-ion levels on the left-hand side; crystal-field levels (with occupation of the one-electron crystal-field levels in parentheses) on the right-hand side. Note that levels belonging to the same subconfiguration tend to run parallel. The energy E and the octahedral crystal field Δ are plotted relative to B, an interelectronic repulsion parameter

Figure 2.8 (d^1 configuration) is the simplest of these three figures. The free ion has fivefold orbital degeneracy (^2D) which is split into two levels (^2E and $^2T_2$) in octahedral symmetry. In the figures the crystal field is taken octahedral, since octahedral coordination is quite common for transition metal ions. The only possible optical absorption transition is from $^2T_2$ to ^2E. This is shown in Fig. 2.11. The energy difference $^2E - {}^2T_2$ is equal to Δ, the crystal field strength. For trivalent transition metal ions Δ is about 20 000 cm^{-1}, i.e. the corresponding optical transition is situated in the visible spectral region. This explains why transition metal ions are usually nicely colored. The $^2T_2 \rightarrow {}^2E$ transition is a clear example of a transition the energy of which is determined by the strength of the crystal field.

Note, however, that this transition is a forbidden one, since it occurs between levels of the d shell. Therefore the parity does not change. Actually the color of transition metal ions is never intense, as far as this type of transitions, also called crystal-field transitions, are involved. The parity selection rule is relaxed by coupling of the electronic transition with vibrations of suitable symmetry [13]. In tetrahedral symmetry, however, the centre of symmetry is lacking, and the parity selection rule is also relaxed in another way, viz. by mixing small amounts of opposite-parity

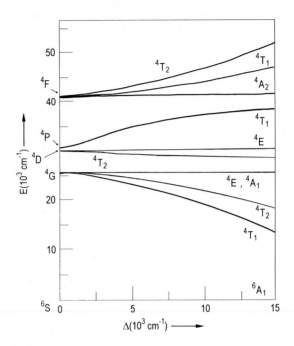

Fig. 2.10. The energy levels of the d^5 configuration as a function of the octahedral crystal field. The abscis is the ground state level ($^6S \rightarrow {}^6A_1$). Only the sextet and the quartets are given. Doublets are omitted for clarity

wave functions into the d wave functions [13]. Actually, the color of tetrahedrally coordinated transition-metal ions is not so weak as that of the octahedrally coordinated ones.

The more-electron cases are considerably more complicated (Figs 2.9 and 2.10). However, remembering the selection rules, we can derive the absorption spectra to be expected. For an ion with three $d$electrons, for example $Cr^{3+}(3d^3)$, the ground level is $^4A_2$. Optical absorption will, in good approximation, occur only to the spin-quartet ($2S + 1 = 4$) levels in view of the spin selection rule. There are only three of these transitions possible, viz. $^4A_2 \rightarrow {}^4T_2$, $^4T_1(^4F)$ and $^4T_1(^4P)$. Indeed the absorption spectrum of the Cr^{3+} ion consists of three absorption bands with low intensity (parity selection rule). This is shown in Fig. 2.12. The spin-forbidden transitions can usually be observed only in very accurate measurements.

The last example is the d^5 configuration, of which Mn^{2+}, used in many luminescent materials, is a well-known representative. The Tanabe-Sugano diagram is given in Fig. 2.10. In octahedral coordination the ground level is $^6A_1$. All optical absorption transitions are parity and spin forbidden. In fact the Mn^{2+} ion is practically colorless.

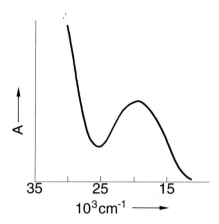

Fig. 2.11. The absorption spectrum of Ti^{3+} ($3d^1$) in aqueous solution. The band at about 20 000 cm^{-1} is the $^2T_2 \rightarrow \ ^2E$ transition; the intense band in the ultraviolet is a charge-transfer transition

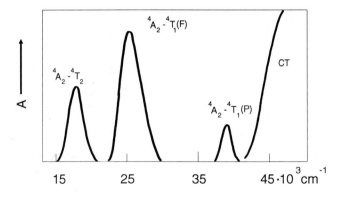

Fig. 2.12. Schematic representation of the absorption spectrum of Cr^{3+} ($3d^3$) in an oxide. The spin-allowed crystal-field transitions are clearly visible. At high energy a charge-transfer (CT) transition occurs

However, Mn^{2+} compounds, like MnF_2 and $MnCl_2$, have a light rose color. The absorption spectrum of MnF_2 is given in Fig. 2.13. A large number of spin-sextet to spin-quartet transitions is observed, in agreement with Fig. 2.10. The molar absorption coefficient is two orders of magnitude lower than for Cr^{3+} in view of the spin selection rule. That for Cr^{3+} is more than three orders of magnitude lower than for allowed transitions (parity selection rule).

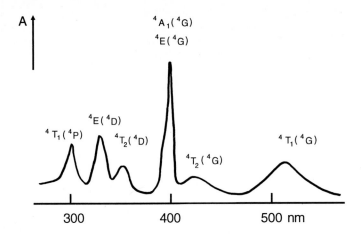

Fig. 2.13. Absorption spectrum of MnF_2

An interesting aspect of the absorption spectrum of Mn^{2+} is the different width of the absorption transitions. Especially the practically coinciding $^6A_1 \rightarrow {}^4A_1$ and ^4E bands are very narrow, and the $^6A_1 \rightarrow {}^4T_1$ and $^4T_2$ rather broad. Above it was shown that the band width is due to coupling with vibrations. Since the crystal field strength varies during the vibration, the Tanabe-Sugano diagrams also predict the width of the absorption bands. If the level reached after absorption runs parallel with the ground level (i.e. the x axis), a variation of Δ will not influence the transition energy, and a narrow absorption band is to be expected. If the excited level has a slope relative to the x axis, a variation of Δ will influence the transition energy, and a broad absorption band is to be expected. The reader can convince himself from the correctness of these statements by comparing Figs 2.11–13 with Figs 2.8–10.

The physical background of these statements was given above. Let us illustrate this for Mn^{2+}. The ground level configuration written in one-electron crystal-field components is $t_2{}^3e^2$. The excited $^4A_1$ and ^4E levels originate from the same configuration. Therefore ΔR is vanishing, and the transitions $^6A_1 \rightarrow {}^4A_1$ and ^4E should appear in the absorption spectrum as lines, in agreement with experiment. However, the levels $^4T_1$ and $^4T_2$ originate mainly from a different configuration, viz. $t_2{}^4e^1$. Since the chemical bond involving a t_2 orbital is different from that involving an e orbital, this configuration has an equilibrium distance which differs from that of the $t_2{}^3e^2$ ground configuration. Therefore, transitions like $^6A_1 \rightarrow {}^4T_1$ and $^4T_2$ involve a change in R_o (i.e. $\Delta R \neq 0$) and should be observed in the spectra as bands with a certain width.

In the ultraviolet region the transition metal ions show usually broad and strong absorption bands due to ligand-to-metal charge-transfer transitions. These will be discussed now for the simple, but important case in which the d electrons are lacking.

This will be indicated here as a d^0 ion. Examples are Cr^{6+} and Mn^{7+} which are strongly colored in oxides (chromates, permanganates). This intense color is due to the fact that the charge-transfer transition has shifted into the visible. From the view point of luminescent materials, the more important examples are V^{5+} ($3d^0$), Nb^{5+} ($4d^0$) and W^{6+} ($5d^0$).

2.3.2 The Transition Metal Ions with d^0 Configuration

Compounds like YVO_4, $YNbO_4$ and $CaWO_4$ are very important materials in view of their luminescence properties. Their absorption spectra show strong and broad bands in the ultraviolet. The transition involved is a charge transfer from oxygen to the d^0 ion [14]. An electron is excited from a non-bonding orbital (on the oxygen ions) to an anti-bonding orbital (mainly d on the metal ion). Therefore the bonding is strongly weakened after optical absorption, so that $\Delta R >> 0$ and the band width is large.

The spectral position of this absorption transition depends on many factors: the ionization potential of the $d^1 \rightarrow d^0$ ionization, the number and nature of the ligands, and the interaction between ions mutually in the lattice. Without further discussion the following examples are given by way of illustration:

- whereas $CaWO_4$ has the first absorption transition at $40\,000$ cm^{-1} (250 nm), $CaMoO_4$ has it at $34\,000$ cm$^-$ (290 nm). This illustrates the higher sixth ionisation potential of Mo (70 eV) relative to W (61 eV), since all other factors are equal.
- whereas $CaWO_4$ with WO_4^{2-} groups has the first absorption transition at $40\,000$ cm^{-1}, that of Ca_3WO_6 with WO_6^{6-} groups is at $35\,000$ cm^{-1}, illustrating the shift of the charge-transfer band to lower energies if the number of ligands increases.
- whereas Ca_3WO_6, with isolated WO_6^{6-} groups, has its charge-transfer band at $35\,000$ cm^{-1}, that of WO_3, with WO_6^{6-} groups sharing oxygen ions, is in the visible (WO_3 is yellow colored); this illustrates the effect of interaction between optical centers with charge-transfer bands.

2.3.3 The Rare Earth Ions ($4f^n$)

The rare earth ions are characterised by an incompletely filled $4f$ shell. The $4f$ orbital lies inside the ion and is shielded from the surroundings by the filled $5s^2$ and $5p^6$ orbitals. Therefore the influence of the host lattice on the optical transitions within the $4f^n$ configuration is small (but essential). Figure 2.14 presents a substantial part of the energy levels originating from the $4f^n$ configuration as a function of n for the trivalent ions. The width of the bars in Fig. 2.14 gives the order of magnitude of the crystal field splitting which is seen to be very small in comparison with the transition metal ions.

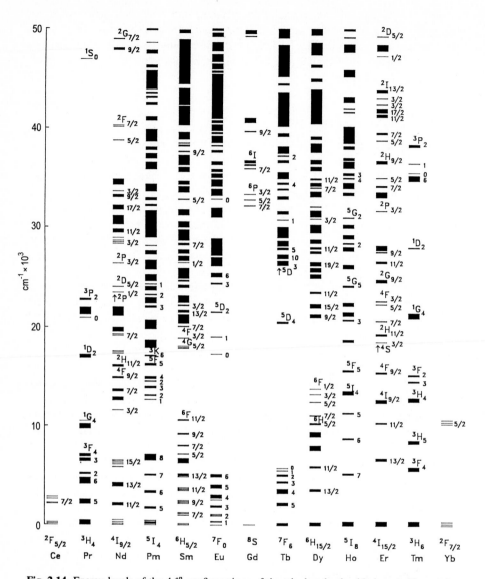

Fig. 2.14. Energy levels of the $4f^n$ configurations of the trivalent lanthanide ions (with permission reproduced from Carnall WT, Goodman GL, Rajnak K, Rana RS (1989) J Chem Phys 90: 343

Optical absorption transitions are strongly forbidden by the parity selection rule. Generally speaking the color of the oxides RE_2O_3 is close to white, although there are energy levels in the visible region. Only Nd_2O_3 (faint violet) is clearly colored. The dark colors of the commercial praseodymium and terbium oxides are due to the simultaneous presence of tri- and tetravalent ions (see below).

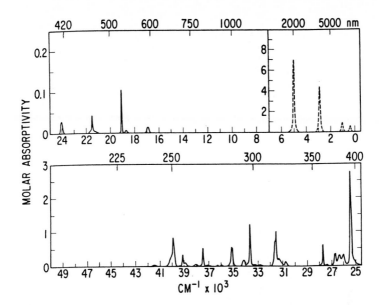

Fig. 2.15. Absorption spectrum of the Eu^{3+} ion in aqueous solution (with permission reproduced from Carnall WT (1979) Handbook on the physics and chemistry of rare earths, vol. 3. North Holland, Amsterdam, p. 171)

Figure 2.15 shows as an example the absorption spectrum of Eu^{3+} in aqueous solution. The sharpness of the lines has already been discussed before: for intraconfigurational $4f^n$ transitions $\Delta R = 0$. Note also the very low values of the molar absorption coefficient.

How is the parity selection rule relaxed? Vibrations have only a very weak influence. For interesting consequences of this influence the reader is referred to Ref. [15]. Of more importance are the uneven components of the crystal-field which are present when the rare earth ion occupies a crystallographic site without inversion symmetry. These uneven components mix a small amount of opposite-parity wave functions (like $5d$) into the $4f$ wavefunctions. In this way the intraconfigurational $4f^n$ transitions obtain at least some intensity. Spectroscopists say it in the following way: the (forbidden) $4f$-$4f$ transition steals some intensity from the (allowed) $4f$-$5d$ transition. The literature contains many treatments of these rare earth spectra, some in a simple way, others in considerable detail [1,16,17,18,19].

If the absorption spectra of the rare earth ions are measured at high enough energy, allowed transitions are also observed as will be discussed now.

2.3.4 The Rare Earth Ions ($4f$-$5d$ and Charge-Transfer Transitions)

The allowed optical transitions of the rare earth ions mentioned above are interconfigurational and consist of two different types, viz.

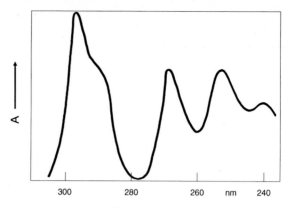

Fig. 2.16. Absorption spectrum of the Ce^{3+} $(4f^1)$ ion in $CaSO_4$. The $4f \rightarrow 5d$ transition has five components due to the crystal-field splitting of the excited $5d$ configuration

- charge-transfer transitions $(4f^n \rightarrow 4f^{n+1}L^{-1}$, where L = ligand)
- $4f^n \rightarrow 4f^{n-1}5d$ transitions.

Both are allowed, both have $\Delta R \neq 0$, and appear in the spectra as broad absorption bands. Charge transfer transitions are found for rare earth ions which like to be reduced, $4f$-$5d$ transitions for ions which like to be oxidized. The tetravalent rare earth ions (Ce^{4+}, Pr^{4+}, Tb^{4+}) show charge-transfer absorption bands [20]. Orange $Y_2O_3 : Tb^{4+}$ has its color due to a charge-transfer absorption band in the visible.

The divalent rare earth ions (Sm^{2+}, Eu^{2+}, Yb^{2+}), on the other hand, show $4f \rightarrow 5d$ transitions, Sm^{2+} in the visible, Eu^{2+} and Yb^{2+} in the long wavelength ultraviolet.

The trivalent ions that have a tendency to become divalent (Sm^{3+}, Eu^{3+}, Yb^{3+}) show charge-transfer absorption bands in the ultraviolet. In the less electronegative sulfides also ions like Nd^{3+}, Dy^{3+}, Ho^{3+}, Er^{3+} and Tm^{3+} show charge-transfer transitions (in the spectral area around $30\,000$ cm^{-1}).

The trivalent ions that have a tendency to become tetravalent (Ce^{3+}, Pr^{3+}, Tb^{3+}) show $4f \rightarrow 5d$ absorption bands in the ultraviolet. An example is given in Fig. 2.16.

This type of optical transition is discussed in more detail in ref. [21].

2.3.5 Ions with s^2 Configuration

Ions with s^2 configuration show strong optical absorption in the ultraviolet due to a $s^2 \rightarrow sp$ transition (parity allowed). The s^2 configuration gives one level for the ground state, viz. $^1S_0$. The subscript indicates the value of J, the total angular momentum of the ion. The sp configuration yields, in sequence of increasing energy, the $^3P_0$, $^3P_1$, $^3P_2$, and $^1P_1$ levels.

In view of the spin selection rule the only optical absorption transition to be expected is $^1S_0 \rightarrow {}^1P_1$. Indeed this transition dominates the absorption spectrum (Fig. 2.17). However, the transition $^1S_0 \rightarrow {}^3P_1$ can also be observed. This is due to spin-orbit coupling which mixes the spin triplet and singlet, the more so for the

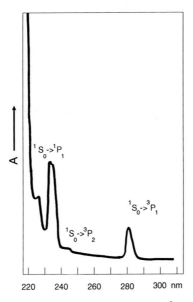

Fig. 2.17. Absorption spectrum of KI-Tl$^+$. The Tl$^+$ ion has $6s^2$ configuration. The transitions originating from the $6s^2 \rightarrow 6s6p$ transition are indicated. The absorption transitions at higher energy are due to charge transfer and the host lattice

heavier elements. Indeed the spin-forbidden transition becomes more intense in the series As$^{3+}(4s^2)$, Sb$^{3+}(5s^2)$, Bi$^{3+}(6s^2)$.

The transition $^1S_0 \rightarrow {}^3P_0$ remains forbidden, since the total angular momentum does not change ($\Delta J = 0$). The $^1S_0 \rightarrow {}^3P_2$ transition remains also forbidden ($\Delta J = 2$ is a forbidden transition), but can obtain some intensity by coupling with vibrations. For more details on the energy levels and transitions of s^2 ions, the reader is referred to Ref. [22].

Since bonding with p electrons is different from that with s electrons, $\Delta R \neq 0$, and broad absorption bands are to be expected. The band width of s^2 ions depends, however, strongly on the host lattice. This will be discussed in the next chapter.

2.3.6 Ions with d^{10} Configuration

Ions with d^{10} configuration like Zn^{2+}, Ga^{3+}, Sb^{5+}, etc. show an intense and broad absorption band in the short wavelength ultraviolet. In recent years it has become clear that such ions can also show luminescence [23]. The exact nature of the optical absorption transition is not clear, but it consists of a charge-transfer transition from the ligands ($2p$ orbital of oxygen) to an antibonding orbital which is situated partly on the d^{10} ion and partly on the ligands. In this way the large width and high intensity of the absorption band can be accounted for.

2.3.7 Other Charge-Transfer Transitions

Several examples of charge-transfer transitions have been given above. These were all of the LMCT (ligand-to-metal charge transfer) type. However, MLCT (metal-to-ligand charge transfer) is also possible, although in oxides not very probable. In coordination compounds these are quite common.

Another type is MMCT (metal-to-metal charge-transfer) [24,25] in which an electron is transferred from one metal ion to another. If the ions involved are of the same element, this is called intervalence charge-transfer. This occurs in Prussian blue which contains Fe^{2+} and Fe^{3+}, and $BaBiO_3$ with Bi^{3+} and Bi^{5+} ions. Examples where the metal ions are different are blue sapphire ($Al_2O_3 : Fe^{2+},Ti^{4+}$; the color is due to $Fe^{2+} \rightarrow Ti^{4+}$ charge transfer), $YVO_4 : Bi^{3+}$ (with $Bi^{3+} \rightarrow V^{5+}$ charge transfer) and the luminescent material $CaWO_4 : Pb^{2+}$ (with $Pb^{2+} \rightarrow W^{6+}$ charge transfer).

As a matter of fact, all absorption bands due to this type of transitions are very broad.

2.3.8 Color Centers

The best-known color center is the F center. It consists of an electron in a halide vacancy in, for example, KCl. It occurs only in solids. In good approximation its absorption spectrum can be described in the same way as that of the hydrogen atom. The first optical transition is $1s \rightarrow 2p$ (parity allowed). In KCl its absorption maximum is in the red, so that a KCl crystal containing F centers is strongly blue colored.

The theory of these transitions is nowadays known in great detail. Many other color centers exist. Their treatment is outside the scope of this book. The reader is referred to Refs [1] and [2].

2.4 Host Lattice Absorption

As we have seen above, absorption of radiation does not necessarily take place in the luminescent center itself, but may also occur in the host lattice. It is obvious to make a simple subdivision into two classes of optical absorption transitions, viz. those which result in free charge carriers (electrons and holes), and those which do not. Photoconductivity measurements can distinguish between these two classes.

An example of the former class is ZnS, the host lattice for cathode-ray phosphors. This compound is a semiconductor. Optical absorption occurs for energies larger than E_g, the width of the forbidden gap. This absorption creates an electron in the conduction band and a hole in the valence band. Since the top of the valence band consists of levels with predominant sulfur character and the bottom of the conduction band of levels with a considerable amount of zinc character, the optical transition is of the charge transfer type. Its position can be shifted by replacing Zn and/or S in ZnS by other elements (see Table 2.3).

Table 2.3. Somesemiconductors of the ZnS type and their optical absorption.

	E_g (eV)	color
ZnS	3.90	white
ZnSe	2.80	orange
ZnTe	2.38	red
CdS	2.58	orange
CdTe	1.59	black

However, not every host lattice yields free electrons and holes upon optical excitation. Ultraviolet irradiation of $CaWO_4$, for example, is absorbed in the WO_4^{2-} groups as described above. In the excited state of the tungstate group the hole (on oxygen) and the electron (on the tungsten) remain together: their interaction energy is strong enough to prevent delocalisation like in ZnS. Such a bound electron-hole pair is called an exciton, and more specifically when the binding is strong, as in $CaWO_4$, a Frenkel exciton [1,26]. Another example is the first absorption band of NaCl, situated at 8eV (155 nm) in the vacuum ultraviolet. It is due to the $3p^6 \rightarrow 3p^54s$ transition on the Cl^- ion. Strongly coupled electron-hole pairs occur only in ionic compounds.

The processes occurring upon irradiating a material with high-energy radiation like cathode rays, X rays or γ rays are rather complicated. After penetration into the solid, this radiation will give ionisation depending on the type and the energy of the radiation. This ionisation creates many secondary electrons. After thermalisation we are left with electron-hole pairs, as we do after irradiation with ultraviolet radiation just over the band gap.

We have described in this chapter the processes and transitions which are responsible for the absorption of radiation, with special attention to ultraviolet radiation and absorption by the center itself. Our systems are now in the excited state. In the following chapters we will consider how they return again to the ground state. Sometimes they follow simply the reverse of the absorption transition, but more often they prefer a different way without hesitating to make large detours.

References

1. Henderson B, Imbusch GF (1989) Optical spectroscopy of inorganic solids. Clarendon, Oxford
2. Stoneham AM (1985) Theory of defects in solids. Clarendon, Oxford.
3 DiBartolo B (1968) Optical interactions in solids. Wiley, New York
4. Atkins PW (1990) Physical chemistry, 4th edn. Oxford University Press, Oxford
5. Jørgensen CK (1962) Absorption spectra and chemical bonding in complexes. Pergamon, Oxford
6. Jørgensen CK (1971) Modern aspects of ligand field theory, North-Holland, Amsterdam
7. Lever ABP (1984) Inorganic electronic spectroscopy, 2nd edn. Elsevier, Amsterdam
8. Duffy JA (1990) Bonding, energy levels and bands in inorganic solids. Longman Scientific and Technical, Harlow
9. Blasse G (1972) J Solid State Chem 4: 52

10. Antic-Fidancev E, Lemaitre-Blaise M, Derouet J, Latourette B, Caro P (1982) C.R. Ac. Sci. Paris 294: 1077
11. Flahaut J (1979) Ch. 31 in Vol. 4 of the Handbook on the physics and chemistry of rare earths, North-Holland, Amsterdam
12. Shriver DF, Atkins PW, Langford CH (1990) Inorganic Chemistry, Oxford University Press, Oxford
13. Cotton FA (1990) Chemical Applications of Group Theory, 3rd edn. Wiley, Chichesterter
14. Blasse G (1980) Structure and Bonding 42: 1
15. Blasse G (1992) Int. Revs Phys. Chem. 11: 71
16. Judd BR (1962) Phys. Rev. 127: 750; Ofelt GS (1962) J. Chem. Phys. 37: 511
17. Carnall WT (1979) Chapter 24 in vol. 3 of the Handbook on the physics and chemistry of rare earths (Gschneidner KA Jr, Eyring L eds) North-Holland Amsterdam
18. Blasse G (1979) Chapter 34 in vol. 4 of the Handbook on the physics and chemistry of rare earths (Gschneidner KA Jr, Eyring L eds) North-Holland Amsterdam; (1987) Spectroscopy of solid-state laser-type materials (DiBartolo B. ed.) Plenum New York (1987) 179
19. Peacock RD (1975) Structure and Bonding 22: 83
20. Hoefdraad HE (1975) J. Inorg. Nucl. Chem. 37: 1917
21. Blasse G (1976) Structure and Bonding 26: 43
22. Ranfagni A, Mugnai D, Bacci M, Viliani G, Fontana MP (1983) Adv. Physics 32: 823
23. Blasse G (1990) Chem. Phys. Letters 175: 237
24. Blasse G (1991) Structure and Bonding 86: 153
25. Brown DB (ed.) (1980) Mixed-valence compounds, Reidel, Dordrecht
26. Kittel C, Introduction to solid state physics, Wiley, New York, several editions.

CHAPTER 3

Radiative Return to the Ground State: Emission

3.1 Introduction

In Chapter 2, several ways were considered in which the luminescent system can absorb the excitation energy. In the following chapters the several possibilities of returning to the ground state are considered. In this chapter we will deal with radiative return to the ground state in the case when the absorption and emission processes occur in the same luminescent center. This situation occurs when photoluminescence is studied on a luminescent center in low concentration in a non-absorbing host lattice (Fig. 1.1).

In Chapter 4 nonradiative return to the ground state will be discussed for a comparable situation. Chapter 5 deals with another possibility of returning to the ground state, viz. transfer of the excitation energy of one ion to another (Fig. 1.3).

This chapter is organized as follows. First a general discussion will be given on the basis of the configurational coordinate diagram (see Chapter 2). Then we will consider a number of different classes of luminescent ions. At the end some special effects, like afterglow and stimulated emission, are treated.

3.2 General Discussion of Emission from a Luminescent Center

Figure 3.1 shows again the configurational coordinate diagram. For the moment we assume that there is an offset between the parabolas of the ground and excited state. According to Chapter 2 absorption occurs in a broad optical band and brings the center in a high vibrational level of the excited state. The center returns first to the lowest vibrational level of the excited state giving up the excess energy to the surroundings. Another way to describe this process is to say that the nuclei adjust their positions to the new (excited) situation, so that the interatomic distances equal the equilibrium distances belonging to the excited state. The configurational coordinate changes by ΔR. This process is called relaxation.

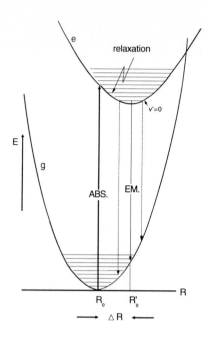

Fig. 3.1. Configurational coordinate diagram (see also Fig. 2.3). The absorption transition g → e is for reasons of clarity drawn as one line only (the transition with maximum intensity). After absorption the system reaches high vibrational levels of the excited state. Subsequently it relaxes to the lowest vibrational level v' = 0 from where emission e → g occurs in a broad band. The parabola offset is given by ΔR

During relaxation there occurs usually no emission, and certainly not of high intensity. This can be easily seen from the rates involved: whereas a very fast emission has a rate of 10^8 s^{-1}, the vibrational rate is about 10^{13} s^{-1}.

From the lowest vibrational level of the excited state the system can return to the ground state spontaneously under emission of radiation. The rules for this process are the same as described for the absorption process. The difference is that emission occurs spontaneously (i.e. in the absence of a radiation field), whereas absorption can only occur when a radiation field is present. The reverse process of absorption is stimulated emission (see Sect. 3.6) and not spontaneous emission.

By emission, the center reaches a high vibrational level of the ground state. Again relaxation occurs, but now to the lowest vibrational level of the ground state. The emission occurs at a lower energy than the absorption due to the relaxation processes (see Fig. 3.1). As an example, Figure 3.2 shows the emission and excitation (= absorption) spectra of the luminescence of Bi^{3+} in LaOCl. The energy difference between the maximum of the (lowest) excitation band and that of the emission band is called the Stokes shift. It will be immediately clear that the larger the value of ΔR is, the larger the Stokes shift and the broader the optical bands involved.

STOKES   SHIFT

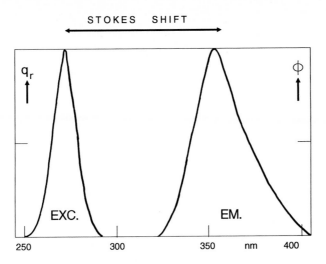

Fig. 3.2. Emission and excitation spectra of the Bi^{3+} luminescence of $LaOCl:Bi^{3+}$. The Stokes shift amounts to about 9000 cm^{-1}. In this and other figures q_r gives the relative quantum output and Φ the spectral radiant power per constant wavelength interval, both in arbitrary units

If the two parabolas have equal force constants (i.e. the same shape), the amount of energy lost in the relaxation process is $Sh\nu$ per parabola, where $h\nu$ is the spacing between two vibrational levels and S an integer. The Stokes shift amounts to $2Sh\nu$. The constant S is called the Huang-Rhys coupling constant. It is proportional to $(\Delta R)^2$ (see e.g. Ref. [1] and measures the strength of the electron-lattice coupling. If S < 1, we are in the weak-coupling regime, if 1 < S < 5, in the intermediate coupling regime, if S > 5, in the strong-coupling regime.

If R denotes the distance between the central metal ion and the ligands, the absorption-emission cycle can be visualized as follows: absorption occurs without change of R, and is followed by expansion of the luminescent center till the new equilibrium distance $(R + \Delta R)$ is reached; subsequently emission occurs without change of distance followed by a contraction ΔR till the equilibrium distance of the ground state is reached. This is a classic description which, however, is often not correct, since the excited state may be distorted relative to the ground state.

As an example of such a distortion we consider the Te^{4+} ion [2]. In a composition like $Cs_2SnCl_6:Te^{4+}$ this ion shows a yellow luminescence (see Fig. 3.3). The value of the Stokes shift is 7000 cm^{-1}. The emission spectrum shows a clear structure (Fig. 3.3) which consists of a large number of equidistant lines. Their energy difference is 240 cm^{-1}. It corresponds to a vibrational mode of the ground state. From Raman spectra it can be derived that this vibrational mode is not the one in which all metal-ligands bonds expand and contract in phase (ν_1). The 240 cm^{-1} mode is a mode in which the luminescent $TeCl_6^{2-}$ octahedron is tetragonally distorted (ν_2) (Fig. 3.4). The configurational coordinate to be taken is ν_2. This means that the $TeCl_6^{2-}$ octahedron is tetragonally distorted during the relaxation after optical absorption. After emission

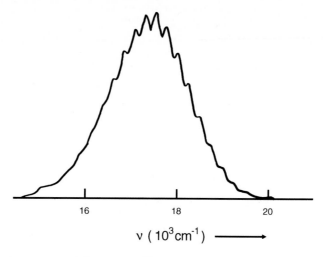

$$\nu \ (\ 10^3 cm^{-1}) \longrightarrow$$

Fig. 3.3. Emission spectrum of $Cs_2SnCl_6 : Te^{4+}$ at 4.2 K

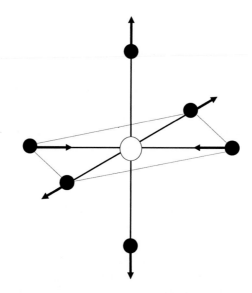

Fig. 3.4. The vibrational ν_2 mode of the octahedron ML_6 (M = metal ion, L : ligand ion)

it relaxes to the undistorted ground state. Similar results have been reported for Cr^{3+} in CrO_6^{9-} and the vanadate group (VO_4^{3-}) (see below).

As a matter of fact the coupling with all possible vibrational modes has to be considered. Fortunately the coupling with only one mode dominates in many cases, i.e. the coupling constant S is small for all modes except this one. Analysis is then

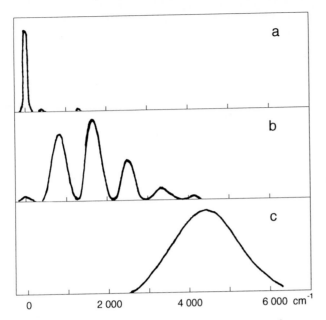

Fig. 3.5. Emission spectra at low temperature of $GdAl_3B_4O_{12}$ (a), UO_2^{2+} (b), and the F centre. The position of the zero-phonon line is always at 0 cm^{-1}. In (a) the zero-phonon line dominates (Gd^{3+}, $S \sim 0$), in (b) it is still observable ($S \sim 2$), and in (c) it has vanished ($S > 5$). The absolute position of the emission is for (a) in the ultraviolet, for (b) in the green, and for (c) in the infrared

possible. If coupling with several modes is effective, the spectral band is broad and the amount of information which can be obtained from it decreases rapidly.

For the emission band shape the same considerations are valid as for the absorption band shape (Chapter 2). Figure 3.5 shows a few emission spectra of different luminescent species, viz. Gd^{3+}, UO_2^{2+} and the F center. These spectra are plotted on the same energy scale and relative to the zero-phonon line. In the case of Gd^{3+} the zero-phonon line dominates (weak coupling), in the case of UO_2^{2+} there is a clear progression in the ν_1 mode (the symmetrical stretching mode; intermediate coupling); in the case of the F center there is a broad band and the intensity of the zero-phonon line has vanished (strong coupling). It is the nature of the luminescent species which determines the value of S (the strength of the electron-lattice coupling).

Nevertheless, there is also an influence of the host lattice. For example, Te^{4+} in Cs_2SnCl_6 shows an emission spectrum in which vibrational structure can be observed (Fig. 3.3). In $ZrP_2O_7 : Te^{4+}$, however, this structure has disappeared and the emission band width has more than doubled. The Bi^{3+} ion shows vibrational structure in its emission band in $ScBO_3$, but not in $LaBO_3$ [3]. The disappearance of the vibrational structure points to an increase of the value of S or of the offset ΔR. It has been shown that a stiff surroundings of the luminescent center restricts the value of ΔR and of S, so that the Stokes shift becomes smaller and vibrational structure may appear in favourable cases. The influence of this stiffness is even more pronounced in the case of processes which enable nonradiative return to the ground state (see Chapter 4).

The influence of the host lattice on the absorption transitions (Chapter 2) is of course also present in the case of emission transitions. This influence is different from the influence of the host lattice on ΔR and S, so that the influence of the host lattice on the emission transition is not easy to unravel: the influence of several effects has to be distinguished, and this is not always easy.

Whereas in absorption spectroscopy the strength of the optical absorption is measured in an easy and classic way [4], this is different in the case of emission spectroscopy. Here the key property is the life time of the excited state. For allowed emission transitions this life time is short, viz. $10^{-7} - 10^{-8}$ s, for strongly forbidden transitions in solids it is much longer, viz. a few 10^{-3} s. This life time is of great importance in many applications. Therefore we discuss this quantity in more detail. For the two-level system of Figure 3.1 (excited state and ground state) the population of the excited state decreases according to

$$\frac{dN_e}{dt} = -N_e P_{eg} \tag{3.1}$$

In Eq. (3.1) the value of N_e gives the number of luminescent ions in the excited state after an excitation pulse, t the time, and P_{eg} the probability for spontaneous emission from the excited to the ground state.

Integration yields

$$N_e(t) = N_e(0)e^{-P_{eg}t} \tag{3.2}$$

which is also written as

$$N_e(t) = N_e(0)e^{-t/\tau_R} \tag{3.3}$$

where $\tau_R (= P_{eg}^{-1})$ is the radiative decay time. A plot of the logarithm of the intensity versus time should give a linear curve. An example is given in Figure 3.6. After a time τ_R the population of the excited state has decreased to $\frac{1}{e}$ (37%).

The expression (2.1) is of course also valid for the emission transition. As for absorption transitions, the nuclear part determines the emission band shape. The electronic part determines the value of the radiative decay time.

The general knowledge discussed in this paragraph will now be illustrated by and applied on several classes of luminescent centers.

3.3 Some Special Classes of Luminescent Centers

3.3.1 Exciton Emission from Alkali Halides

This section starts with the alkali halides, because the intrinsic luminescent center in these compounds shows a complicated relaxation in the excited state which has

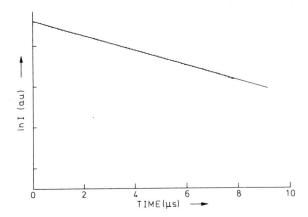

Fig. 3.6. Decay curve of the Eu^{2+} emission of $SrB_4O_7 : Eu^{2+}$ at room temperature. The luminescence intensity I is plotted logarithmically versus the time after the excitation pulse. In agreement with Eq. (3.3) a straight line is found

recently been unravelled by modern experimental techniques. This example illustrates the process of relaxation, and simultaneously the power of the instrumental technique used, viz. femtosecond spectroscopy [5].

Consider as a specific example KCl, a very simple compound indeed. In Chapter 2, its lowest optical absorption band was mentioned to be due to the $3p^6 \rightarrow 3p^54s$ transition on the Cl^- ion. The excited state can be considered as a hole on the Cl^- ion (in the $3p$ shell) and an electron in the direct neighbourhood of the Cl^- ion, since the outer $4s$ orbital spreads over the K^+ ions. Now we consider what happens after the absorption process. The hole prefers to bind two Cl^- ions forming a V_K centre: this centre consists of a Cl_2^- pseudomolecule on the site of two Cl^- ions in the lattice. The electron circles around the V_K centre. In this way a self-trapped exciton is formed. An exciton is a state consisting of an electron and a hole bound together. By the relaxation process ($Cl^{-*} \rightarrow V_K.e$) the exciton has lowered its energy and is now trapped in the lattice.

Up till a few years ago it was generally assumed that the luminescence of KCl was due to self-trapped exciton recombination, i.e. the electron falls in the hole of the V_K center and the energy of the exciton is emitted as radiation. Recently, however, it was shown that the $V_K.e$ exciton can relax further: the Cl_2^- pseudomolecule moves to the lattice site of one Cl^- ion (this is called an H center), the electron to the other Cl^- site which is now vacant (this is the well-known F center). The new relaxed state is an F.H pair. It has a lower energy than the $V_K.e$ relaxed state. The several steps in the relaxation process are depicted in Fig. 3.7. As a matter of fact such a large relaxation results in an enormous Stokes shift (several eV's for alkali halides). After emission, the F.H configuration relaxes back to the ground state, i.e. $e_{Cl}.(Cl_2^-)_{Cl} \rightarrow$ $2Cl dCl$-. The subscript indicates the lattice site.

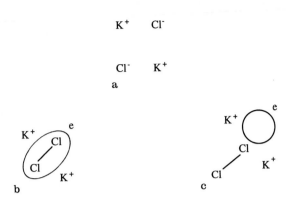

Fig. 3.7. Schematic representation of relaxed excited states in an alkali halide. *a*: ground state; *b*: self-trapped exciton consisting of a V_K centre and an electon; *c*: F.H pair centre. The electron is represented by its orbit (drawn line) marked by the letter *e*, the Cl_2^- pseudomolecule (i.e. the trapped hole) by Cl–Cl. See also text

This example shows clearly that the emission process is very different from the (simple) absorption process. For all details the reader is referred to the literature [5]. Finally we draw attention to the fact that the life time of the relaxed self-trapped exciton in the alkali halides is longer ($\sim 10^{-6}$s) than expected for an allowed transition ($10^{-7} - 10^{-8}$s). This is ascribed to the fact that the emitting state contains an amount of spin triplet character. Such a triplet state arises when the spins of the electron and the hole are oriented parallel. The emission transition becomes (partly) forbidden by the spin selection rule (see Chapter 2).

3.3.2 Rare Earth Ions (Line Emission)

The energy levels of the trivalent rare earth ions which arise from the $4f^n$ configuration are given in Figure 2.14. In a configurational coordinate diagram these levels appear as parallel parabolas ($\Delta R = 0$), because the $4f$ electrons are well shielded from the surroundings. Emission transitions yield, therefore, sharp lines in the spectra. Since the parity does not change in such a transition, the life time of the excited state is long ($\sim 10^{-3}$s).

The energy levels presented in Figure 2.14 are actually split by the crystal field. As a matter of fact the splitting is very small due to the shielding by the $5s^2$ and $5p^6$ electrons: whereas the crystal field strength in case of transition metal ions (d^n) is characteristically a few times $10\,000$ cm^{-1}, it amounts in the rare earth ions (f^n) a few times 100 cm^{-1}.

After these general considerations some special cases will be dealt with.

a. Gd^{3+} $(4f^7)$

This ion has a half-filled $4f$ shell which gives a very stable ${}^8S_{7/2}$ ground state. The excited levels are at energies higher than $32\,000$ cm^{-1}. As a consequence the emission of Gd^{3+} is in the ultraviolet spectral region. The ${}^8S_{7/2}$ level (orbitally nondegenerate) cannot be split by the crystal field. This limits the low-temperature emission spectrum to one line, viz. from the lowest crystal field level of the ${}^6P_{7/2}$ level to ${}^8S_{7/2}$. However, usually the real spectrum consists of more than one line for several reasons.

 – in the first place one usually observes weak vibronic transitions at energies below that of the electronic ${}^6P_{7/2} \to {}^8S_{7/2}$ transition. In the vibronic transition two
   transitions occur simultaneously, viz. the electronic one (in this case ${}^6P_{7/2} \to$
   ${}^8S_{7/2}$) and a vibrational one. Therefore, the energy difference between the electronic transition (in this context often called zero-phonon transition or origin) and
   a vibronic one yields the vibrational frequency of the mode which is excited in
   the emission process. In Figure 3.5 a vibronic line occurs in the Gd^{3+} emission
   spectrum at 1350 cm^{-1} below the electronic origin, which shows that vibrations
   of the borate group are involved. These vibronic transitions are observed in many
   rare-earth spectra (see, for example, Ref. [6].
 – in the second place one may observe at higher energy than the electronic transition
   other transitions which are due to transitions from the higher crystal field levels
   of the ${}^6P_{7/2}$ level. Since the crystal field splitting is small, even at 4.2 K their
   population may be sizable. In Figure 3.8 we show as an example the emission
   spectrum of Gd^{3+} in $LuTaO_4$ at room temperature: there are four lines in the
   ${}^6P_{7/2} \to {}^8S_{7/2}$ transition originating from the four crystal field levels of ${}^6P_{7/2}$,
   and, in addition, three lines belonging to the ${}^6P_{5/2} \to {}^8S_{7/2}$ level which is also
   thermally populated at room temperature.
 – if one excites with high enough energy (e.g. X rays) many more transitions are
   observed. An example is given in Figure 3.9. The composition ($LaF_3 : Gd^{3+}$)
   shows under X-ray excitation, in sequence of increasing energy, emission from
   the ${}^6P$, ${}^6I$, ${}^6D$, and ${}^6G$ levels [7].

b. Eu^{3+} $(4f^6)$

The emission of this ion consists usually of lines in the red spectral area. These lines have found an important application in lighting and display (color television). These lines correspond to transitions from the excited ${}^5D_0$ level to the ${}^7F_J$ ($J = 0, 1, 2, 3, 4, 5, 6$) levels of the $4f^6$ configuration. Since the ${}^5D_0$ level will not be split by the crystal field (because $J = 0$), the splitting of the emission transitions yields the crystal-field splitting of the ${}^7F_J$ levels. This is illustrated in Figure 3.10. In addition to these emission lines one observes often also emission from higher ${}^5D$ levels, viz. ${}^5D_1$, ${}^5D_2$ and even ${}^5D_3$. The factors determining their presence or absence will be discussed in Chapter 4.

The ${}^5D_0-{}^7F_J$ emission is very suitable to survey the transition probabilities of the sharp spectral features of the rare earths. If a rare-earth ion occupies in the crystal lattice a site with inversion symmetry, optical transitions between levels of the $4f^n$

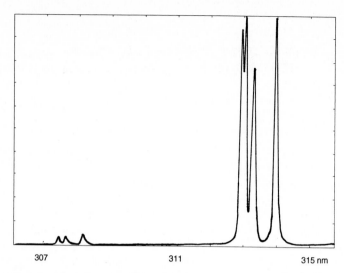

Fig. 3.8. Emission spectrum of the Gd^{3+} luminescence of $LuTaO_4{:}Gd^{3+}$ at room temperature. The $^6P_{7/2} \rightarrow {}^8S_{7/2}$ transition shows four components (longer wavelength side), the $^6P_{5/2} \rightarrow {}^8S_{7/2}$ transition three (shorter wavelength side)

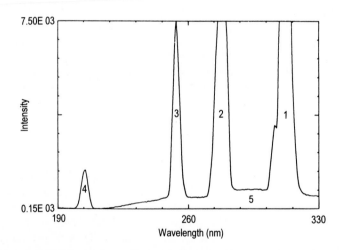

Fig. 3.9. The emission spectrum of the X-ray excited Gd^{3+} luminescence of $LaF_3 : Gd^{3+}$. Line 1 is the $^6P \rightarrow {}^8S$ transition, line 2 $^6I \rightarrow {}^8S$, line 3 $^6D \rightarrow {}^8S$, line 4 $^6G \rightarrow {}^8S$, whereas band 5 is the self-trapped exciton emission of LaF_3 (V_K type, see Sect. 3.3.1)

configuration are strictly forbidden as electric-dipole transition (parity selection rule). They can only occur as (the much weaker) magnetic-dipole transitions which obey the

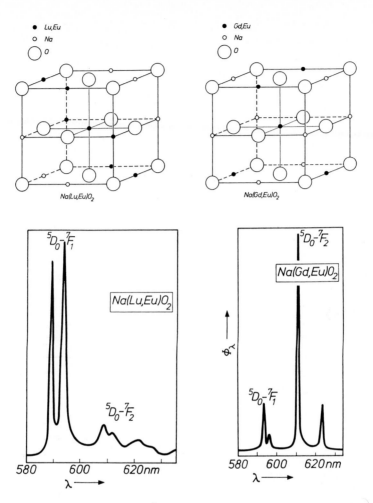

Fig. 3.10. The emission spectrum of Eu^{3+} in $NaLuO_2$ and $NaGdO_2$. In the $NaLuO_2$: Eu^{3+} spectrum the $^5D_0-^7F_1$ lines dominate, in the $NaGdO_2$: Eu^{3+} spectrum the $^5D_0-^7F_2$ lines. At the top of the Fig., a schematical presentation of the crystal structures of the host lattices is given. See also text

selection rule $\Delta J = 0, \pm1$ (but $J = 0$ to $J = 0$ forbidden) or as vibronic electric-dipole transitions.

If there is no inversion symmetry at the site of the rare-earth ion, the uneven crystal field components can mix opposite-parity states into the $4f^n$-configurational levels (Sect. 2.3.3). The electric-dipole transitions are now no longer strictly forbidden and appear as (weak) lines in the spectra, the so-called forced electric-dipole transitions. Some transitions, viz. those with $\Delta J = 0, \pm2$, are hypersensitive to this effect. Even for small deviations from inversion symmetry, they appear dominantly in the spectrum.

We consider again Figure 3.10 for an illustration of these statements. The figure shows the emission spectra of $NaLuO_2 : Eu^{3+}$ and $NaGdO_2 : Eu^{3+}$. Both host lattices have the rocksalt structure, but with a different superstructure between mono- and trivalent metal ions. In $NaLuO_2$ the rare earth ions occupy a site with inversion symmetry. In $NaGdO_2$ the rare earth ions are octahedrally coordinated but due to the superstructure there is a small deviation from inversion symmetry.

In $NaLuO_2 : Eu^{3+}$ the $^5D_0-^7F_1$ emission transition is dominating. All other transitions occur only as very weak and broad lines. These are the vibronic transitions; the electronic origins are lacking. Since the starting level is $^5D_0$, the only possible magnetic-dipole transition is $^5D_0-^7F_1$, as observed experimentally. The trigonal crystal field at the rare earth site in $NaLuO_2$ splits this transition into two lines.

In $NaGdO_2 : Eu^{3+}$ the $^5D_0-^7F_2$ emission transition dominates, but other lines are also present. The Eu^{3+} case is so illustrative, because the theory of forced electric-dipole transitions [8] yields a selection rule in case the initial level has $J = 0$. Transitions to levels with uneven J are forbidden. Further $J = 0 \rightarrow J = 0$ is forbidden, because the total orbital momentum does not change. This restricts the spectrum to: $^5D_0-^7F_1$, present as magnetic-dipole emission, but overruled by the forced electric-dipole emission,
$^5D_0-^7F_2$, a hypersensitive forced electric-dipole emission, which indeed is dominating,
$^5D_0-^7F_{4,6}$, weak forced electric-dipole emissions.

For applications it is required that the main emission is concentrated in the $^5D_0-^7F_2$ transition. This illustrates the importance of hypersensitivity in materials research.

c. Tb^{3+} $(4f^8)$

The emission of Tb^{3+} is due to transitions $^5D_4-^7F_J$ which are mainly in the green. Often there is a considerable contribution to the emission from the higher-level emission $^5D_3-^7F_J$, mainly in the blue. Figure 3.11 gives an example of a Tb^{3+} emission spectrum. Since the J values, involved in the transitions, are high, the crystal field splits the levels in many sublevels which gives the spectrum its complicated appearance.

d. Sm^{3+} $(4f^5)$

The emission of Sm^{3+} is situated in the orange-red spectral region and consists of transitions from the $^4G_{5/2}$ level to the ground state $^6H_{5/2}$ and higher levels ^6H_J ($J > \frac{5}{2}$).

e. Dy^{3+} $(4f^9)$

The emission of Dy^{3+} originates from the $^4F_{9/2}$ level. Dominating are the transitions to $^6H_{15/2}$ ($\sim$ 470 nm) and $^6H_{13/2}$ ($\sim$ 570 nm). The latter one has $\Delta J = 2$ and is hypersensitive. The emission has a whitish color which turns to yellow in host lattices where hypersensitivity is pronounced.

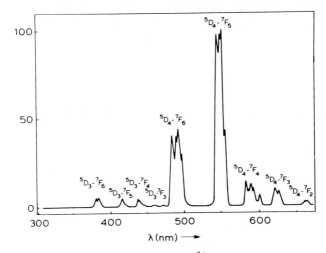

Fig. 3.11. The Tb^{3+} emission spectrum of $GdTaO_4 : Tb^{3+}$

f. Pr^{3+} $(4f^2)$

The emission color of Pr^{3+} depends strongly on the host lattice. If the emission originates from the $^3P_0$ level, it may be green ($^3P_0$–$^3H_4$) like in Gd_2O_2S : Pr, but red lines may also be strong ($^3P_0$–$^3H_6$, $^3F_2$) like in $LiYF_4$: Pr. If the emission originates from the $^1D_2$ level, it is in the red and near-infrared. The factors determining whether emission occurs from $^3P_0$ or $^1D_2$ will be discussed later. The decay time of the $^3P_0$ emission is short for a rare earth ion (tens of μs). Not only is there no spin selection rule active, but also the $4f$ orbitals are probably more spread out in the lighter rare-earths (with lower nuclear charge) facilitating the mixing with opposite-parity states [9].

Rare-earth ion emission is not necessarily sharp line emission, as we will see now.

3.3.3 Rare Earth Ions (Band Emission)

Several rare earth ions show broad band emission. In this emission transition we are dealing with an electron which returns from a $5d$ orbital to the $4f$ orbital (see also Sect. 2.3.4). First we discuss trivalent ions (Ce^{3+}, Pr^{3+}, Nd^{3+}), later divalent ions (Eu^{2+}, Sm^{2+}, Yb^{2+}).

a. Trivalent Ions

The Ce^{3+} ion $(4f^1)$ is the most simple example, since it is a one-electron case. The excited configuration is $5d^1$. The $4f^1$ ground state configuration yields two levels, viz. $^2F_{5/2}$ and $^2F_{7/2}$, separated by some 2000 cm^{-1} due to spin-orbit coupling. The

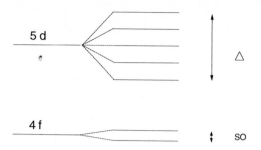

Fig. 3.12. The simplified energy level scheme of the Ce^{3+} ion ($4f^1$). On the left hand side only the $4f$level and the $5d$level are given without taking into account further interactions. On the right hand side the spin-orbit (SO) coupling has split the $4f$level into two components (about 2000 cm^{-1} apart), and the crystal field (Δ) has split the $5d$level into five crystal-field components spanning together some 15 000 cm^{-1}

$5d^1$ configuration is split by the crystal field in 2 to 5 components. The total splitting amounts to some 15 000 cm^{-1} (Fig. 3.12).

The emission occurs from the lowest crystal field component of the $5d^1$ configuration to the two levels of the ground state. This gives the Ce^{3+} emission its typical double-band shape (Fig. 3.13). Since the $5d \rightarrow 4f$transition is parity allowed and spin selection is not appropriate, the emission transition is a fully allowed one. The decay time of the Ce^{3+} emission is short, viz. a few ten ns. The decay time is longer if the emission is at longer wavelengths: 20 ns for the 300 nm emission of CeF_3, and 70 ns for the 550 nm emission of $Y_3Al_5O_{12} : Ce^{3+}$. It can be derived that for a given transition the decay time τ is proportional to the square of the emission wavelength λ [10]: $\tau \sim \lambda^2$.

The Stokes shift of the Ce^{3+} emission is never very large and varies from a thousand to a few thousand wave numbers (medium coupling case). The spectral position of the emission band depends on three factors:

- covalency (the nephelauxetic effect) which will reduce the energy difference between the $4f^1$ and $5d^1$ configurations
- crystal field splitting of the $5d^1$ configuration: a large low-symmetry crystal field will lower the lowest crystal-field component from which the emission originates.
- the Stokes shift.

Usually the Ce^{3+} emission is in the ultraviolet or blue spectral region, but in $Y_3Al_5O_{12}$ it is in the green and red (crystal-field effect), and in CaS in the red (covalency effect).

Under certain conditions $5d-4f$emission has also been observed for Pr^{3+} ($4f^2$) and Nd^{3+} ($4f^3$). For example, $LaB_3O_6 : Pr^{3+}$ shows band emission around 260 nm and $LaF_3 : Nd^{3+}$ around 175 nm. Due to the $\tau \sim \lambda^2$ relation, the decay time of the latter is only 6 ns [11]. However, these ions have an alternative way to emit, viz. by an emission transition in the $4f^n$ configuration.

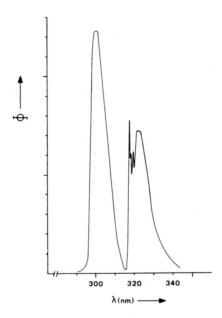

Fig. 3.13. The Ce^{3+} emission spectrum of $LiYF_4:Ce^{3+}$ at 4.2 K. The two bands correspond to the transition from the lowest $5d$ crystal-field component to the two components of the $4f$ ground state. The longer wavelength component shows vibrational structure

b. Divalent Ions

In this group the most well-known and widely applied example is the Eu^{2+} $(4f^7)$ ion which shows a $5d \rightarrow 4f$ emission which can vary from long-wavelength ultraviolet to yellow. Its decay time is about 1 μs. This is due to the fact that the emitting level contains (spin) octets and sextets, whereas the ground state level (^8S from $4f^7$) is an octet, so that the spin selection rule slows down the optical transition rate.

The host lattice dependence of the emission colour of the Eu^{2+} ion is determined by the same factors as in the case of the Ce^{3+} ion. If the crystal field is weak and the amount of covalency low, the lowest component of the $4f^6 5d$ configuration of the Eu^{2+} ion may shift to such high energy, that the $^6P_{7/2}$ level of the $4f^7$ configuration lies below it. At low temperatures sharp-line emission due to the $^6P_{7/2} \rightarrow {}^8S_{7/2}$ transition occurs. This has been observed for quite a number of Eu^{2+}-activated compounds. As an example we can mention $SrB_4O_7:Eu^{2+}$ [12].

Figure 3.14 gives the emission spectrum as a function of temperature. At 4.2 K there is line emission from $^6P_{7/2}$ (and a weak vibronic structure). At 35 K the thermally activated emission from the higher crystal-field components of $^6P_{7/2}$ appears, together with a broad band due to the $4f^6 5d \rightarrow 4f^7$ transition. This band has a zero-phonon line, indicated 0. At 110 K the band dominates.

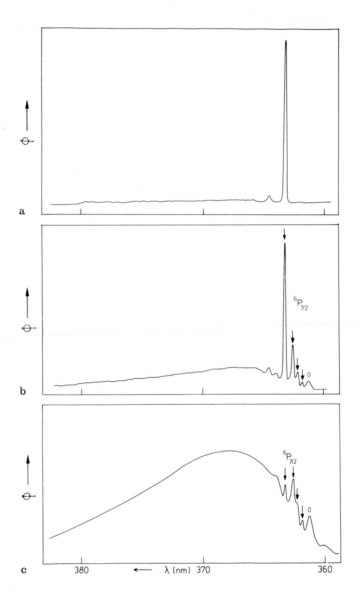

Fig. 3.14. The Eu^{2+} emission spectrum of SrB$_4$O$_7$: Eu^{2+} as a function of temperature. See also text. *a*: 4.2 K (^{6}P$_{7/2}$ → ^{8}S$_{7/2}$ line emission); *b*: 35 K (line emissions, and broad band emission due to the $4f^65d$ → $4f^7$ transition); *c*: 110 K (band emission dominates)

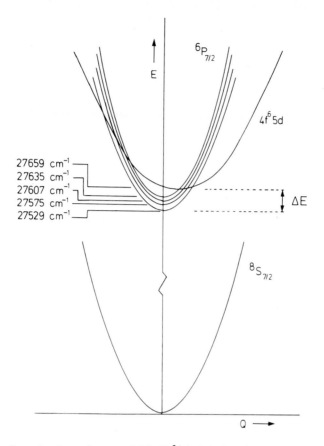

Fig. 3.15. Configurational coordinate model for Eu^{2+} in SrB_4O_7. Figs. 3.14 and 3.15 are derived from A. Meijerink, thesis, University Utrecht, 1990

Figure 3.15 gives the relevant configurational coordinate diagram with the four $^6P_{7/2}$ crystal-field levels and the lowest component of the $4f^65d$ configuration (with a different equilibrium distance).

Finally Figure 3.16 gives the decay time of the Eu^{2+} emission in SrB_4O_7. At low temperatures it is 440 μs (parity forbidden $^6P \rightarrow {}^8S$ transition), but at higher temperatures it decreases rapidly due to the occurrence of the faster $5d \rightarrow 4f$ emission.

The Sm^{2+} ion ($4f^6$) can show $5d \rightarrow 4f$ emission in the red. However, if the lowest level of the $4f^55d$ configuration is at high energy, the intraconfigurational $4f^6$ emission is observed. This runs parallel with the case of Eu^{3+}, although the Sm^{2+} transitions are at a much longer wavelength.

The Yb^{2+} ($4f^{14}$) ion can only show one emission, viz. $4f^{13}5d \rightarrow 4f^{14}$. It is observed in the ultraviolet or blue. An example is given in Fig. 3.17. In the case of this ion the spin selection rule is of even more importance, since the observed decay times are very long for a $5d \rightarrow 4f$ transition (a few ms, Ref. [13]).

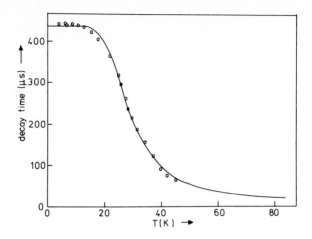

Fig. 3.16. Decay time of the Eu^{2+} emission of $SrB_4O_7 : Eu^{2+}$ as a function of temperature

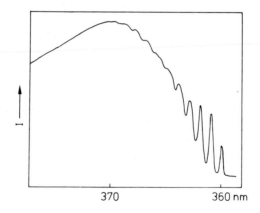

Fig. 3.17. The Yb^{2+} emission spectrum of $SrB_4O_7 : Yb^{2+}$ at 4.2 K

3.3.4 Transition Metal Ions

The luminescence of the transition metal ions will be discussed using the Tanabe-Sugano diagrams (Sect. 2.3.1). First we will consider ions which have played or still play an important role in luminescent materials, viz. Cr^{3+} and Mn^{4+} with d^3 configuration and Mn^{2+} with d^5 configuration. Then we will mention some ions which became more recently of interest. For a detailed account of this field the reader is referred to ref. [1].

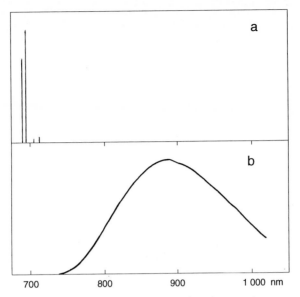

Fig. 3.18. Emission spectra of Cr^{3+}. *a*: $Al_2O_3 : Cr^{3+}$ ($^2E \rightarrow {}^4A_2$ line emission); *b*: $Mg_4Nb_2O_9 : Cr^{3+}$ ($^4T_2 \rightarrow {}^4A_2$ band emission)

a. Cr^{3+} (d^3)

The luminescence of Cr^{3+} in Al_2O_3 (ruby) has already been mentioned in Chapter 1. It formed the basis of the first solid state laser in 1960. This emission consists of two sharp lines (the so-called R lines) in the far red (see Fig. 3.18). Since it is a line, it must be due to the transition $^2E \rightarrow {}^4A_2$ (Fig. 2.9); generally speaking the emission of transition metal ions originates from the lowest excited state. The life time of the excited state amounts to some ms, because the parity selection rule as well as the spin selection rule apply. The emission line is followed by some weak vibronic transitions: obviously this emission transition belongs to the weak-coupling case.

Not always the ^2E level is the lowest excited state. For relatively low crystal fields the $^4T_2$ level is lower. In that case the emission changes character. It is now broad-band $^4T_2$–$^4A_2$ emission in the infrared with a decay time of ~ 100 μs. As an example, Fig. 3.18 gives the emission spectrum of $Mg_4Nb_2O_9 : Cr^{3+}$. This emission forms the basis of tunable infrared lasers. Figure 3.18 also illustrates how dramatic the crystal field influences the emission spectrum in the case of d^3 ions.

In $YAl_3B_4O_{12} : Cr^{3+}$, the emission at 4.2 K is $^2E \rightarrow {}^4A_2$, but at higher temperatures $^4T_2 \rightarrow {}^4A_2$. Although the ^2E level is below $^4T_2$, the latter can be thermally occupied at higher temperatures. Since the $^4T_2 \rightarrow {}^4A_2$ transition probability is higher than that of the $^2E \rightarrow {}^4A_2$ transition due to the spin selection rule, the $^4T_2$ emission rapidly dominates the emission spectrum.

The strength of the crystal field on the Cr^{3+} ion is therefore of imperative importance for its optical properties. Its color is red for high crystal field strength (ruby), and green for low crystal field strength. Its emission is a narrow line in the red in the former case, but a broad band in the near infrared in the latter. In some cases the $^4T_2 \rightarrow {}^4A_2$ emission band shows vibrational structure. There is not only a progression in the symmetric stretching mode ν_1, but also in ν_2 (see Fig. 3.4) indicating a tetragonally distorted excited state.

b. $Mn^{4+}(d^3)$

This ion is isoelectronic with Cr^{3+}, but the crystal field at the higher charged Mn^{4+} ion is stronger, so that the Mn^{4+} emission is always $^2E \rightarrow {}^4A_2$. Usually the vibronics are more intense than for Cr^{3+} [14].

c. $Mn^{2+}(d^5)$

The Mn^{2+} ion has an emission which consists of a broad band, the position of which depends strongly on the host lattice. The emission can vary from green to deep red. The decay time of this emission is of the order of ms. From the Tanabe-Sugano diagram (Fig. 2.10) we derive that the emission corresponds to the $^4T_1 \rightarrow {}^6A_1$ transition. This explains all the spectral properties: a broad band due to different slopes of the energy levels, a long decay time due to the spin selection rule, and a dependence of the emission color on the host lattice due to the dependence on crystal field. Tetrahedrally coordinated Mn^{2+} (weak crystal-field) usually gives a green emission, octahedrally coordinated Mn^{2+} (stronger crystal field) gives an orange to red emission.

d. Other d^n ions

The Ti^{3+} ion ($3d^1$) gives a broad-band emission in the near infrared due to the $^2E \rightarrow {}^2T_2$ transition. The titanium-sapphire laser is based on this emission.

The Ni^{2+} ion ($3d^8$) gives a complicated emission spectrum due to the appearance of emission transitions from more than one level. Luminescence from $KMgF_3 : Ni^{2+}$ appears in the near infrared ($^3T_2 \rightarrow {}^3A_2$), the red ($^1T_2 \rightarrow {}^3T_2$) and the green ($^1T_2 \rightarrow {}^3A_2$).

Güdel and coworkers, in recent years, have reported many near infrared emissions from transition metal ions with "unusual" valencies ($V^{2+}(3d^3)$, $V^{3+}(3d^2)$, $Ti^{2+}(3d^2)$, $Mn^{5+}(3d^2)$ [15]).

3.3.5 d^0 Complex Ions

Complexes of transition metal ions with a formally empty d shell show often intense broad-band emission with a large Stokes shift (10 000–20 000 cm^{-1}). Examples are VO_4^{3-}, NbO_6^{7-}, WO_4^{2-} and WO_6^{6-} [16]. The excited state is considered to be a charge-transfer state, i.e. electronic charge has moved from the oxygen ligands to the central metal ion. The real amount of charge transfer is usually small, but a considerable amount of electronic reorganisation occurs, in which electrons are promoted from

Table 3.1. Decay times τ of some luminescent compounds with d^0 metal ions at 4.2 K.

Compound	τ (μs)
YVO_4	500
$KVOF_4$	33500
$Mg_4Nb_2O_9$	100
$CsNbOP_2O_7$	500
$CaMoO_4$	250
$CaWO_4$	330

bonding orbitals (in the ground state) to antibonding orbitals (in the excited state). The value of ΔR is large, the Stokes shift is large, and the spectral bands broad.

Especially the complexes with the lighter metal ions show long decay times of their emission. Table 3.1 gives some examples. Following early suggestions [16], Van der Waals et al. were able to prove that the emitting state is a spin triplet [17]. They showed also that the excited state is strongly distorted due to the Jahn-Teller effect. Here we meet another clear example of an excited state which is distorted relative to the ground state.

Octahedral complexes of this type have a smaller Stokes shift than the tetrahedral ones. The important consequences of this will be outlined in Chapter 5. Although not understood, certain structural configurations seem to promote efficient luminescence, for example, edge or face sharing of octahedral complexes (Li_3NbO_4, $Ba_3W_2O_9$), and the occurrence of one short metal-oxygen distance ($CsNbOP_2O_7$, $Ba_2TiOSi_2O_7$, $KVOF_4$ and vanadate on silica [18]).

Although it was believed for years that the emission spectra of these species were fully structureless, in recent years vibrational structure has been reported for several cases. A beautiful example is given in Fig. 3.19. This relates to vanadate on a silica surface. The progression is in a vibrational mode with a frequency of 950 cm^{-1}. This is the stretching vibration between vanadium and the oxygen pointing out of the silica surface.

The presence of ions with low-lying energy levels, for example ions with s^2 configuration, influence the d^0-complex luminescence drastically. For example, $CaWO_4 : Pb^{2+}(6s^2)$ shifts its emission to longer wavelength relative to undoped $CaWO_4$ and the quenching temperature goes up. A nice example is $YVO_4 : Bi^{3+}$ ($6s^2$) which has a yellow emission, whereas YVO_4 has blue emission. Arguments have been given to ascribe the new emitting states to charge-transfer states in which the $6s$ electron is transferred to the empty $d$orbital (metal-to-metal charge transfer [19]).

3.3.6 d^{10} Ions

The emission transitions of ions with d^{10} configuration are of a complicated nature and are only partly understood. For clarity these ions are here divided into two classes,

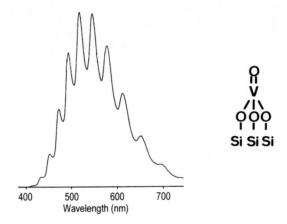

Fig. 3.19. Emission spectrum of the vanadate group on silica at 4.2 K

viz. the monovalent ones (Cu^+ and Ag^+) and the higher valent ones (for example Zn^{2+}, Ga^{3+}, Sb^{5+}, Te^{6+}).

a. Monovalent Ions

Complexes with monovalent d^{10} ions show often efficient emission at room temperature. For Cu^+ the reader can find a summary in Ref. [19]. The emission transition has been assigned to a $d^9s \to d^{10}$ transition, a ligand-to-metal charge-transfer transition, or a metal-to-ligand charge-transfer transition, depending on the ligands. In the mean time the first assignment has been put in doubt, since optically detected electron paramagnetic resonance measurements on the excited state of Cu^+ in NaF point to a very low spin density in the Cu $4s$ orbital [20]. As an alternative, the excited state may be thought of as a Cu^{2+} ion which distorts its surroundings due to the Jahn-Teller effect and an electron which has moved away from the hole in the d shell so that an exciton state is formed. This would be another example of an impressive relaxation after the absorption process.

The Stokes shift of the Cu^+ emission is usually large (≥ 5000 cm^{-1}), indicating the strong-coupling scheme. About the Ag^+ ion less is known, but what is known shows a similarity with the Cu^+ data.

b. Higher Valent Ions

Luminescence from higher valent ions with d^{10} configuration has been questioned for a long time. Nowadays there exists strong evidence for such a luminescence. Examples are $Zn_4O(BO_2)_6$, $LiGaO_2$, $KSiSbO_5$ and $LiZrTeO_6$. The Stokes shifts are very large. Table 3.2 gives some spectral data [21].

The nature of this emission is not yet completely clear, but probably it is a charge-transfer transition, LMCT in absorption. However, also a transition on the oxygen ion ($2p^6 \to 2p^53s$) plays a role. Such an interpretation implies indeed a large

Table 3.2. Some data on the luminescence of complexes with a central d^{10} metal ion. All values at 4.2 K and in cm^{-1}.

Compound	Complex	Emission max.	Excitation max.	Stokes shift
$Zn_4B_6O_{13}$	$Zn(II)O_4$	22.000	40.000	18.000
$LiGaO_2$	$Ga(III)O_4$	27.000	45.000	18.000
$KSbSiO_5$	$Sb(V)O_6$	21.000	41.500	20.500
Li_2ZrTeO_6	$Te(VI)O_6$	16.000	33.000	17.000

Table 3.3. Stokes shift of the Bi^{3+} emission in several host lattices [23].

Composition	Stokes shift (cm^{-1})
$Cs_2NaYCl_6 : Bi$	800
$ScBO_3 : Bi$	1800
$YAl_3B_4O_{12} : Bi$	2700
$CaLaAlO_4 : Bi$	7700
$LaOCl : Bi$	8500
$La_2O_3 : Bi$	10 800
$Bi_2Al_4O_9$	16 000
$Bi_4Ge_3O_{12}$	17 600
$LaPO_4 : Bi$	19 200
$Bi_2Ge_3O_9$	20 000

amount of relaxation in the excited state and has been confirmed by molecular-orbital calculations.

Indeed the O^{2-} ion, when isolated in the lattice, can yield an emission of the type $2p^53s \rightarrow 2p^6$. Examples are $LiF : O^{2-}$, $CdF_2 : O^{2-}$ and $SrLa_2OBeO_4$. Since the $3s$ electron in the excited state has a widely spread orbital, this transition on oxygen, charge transfer in d^0 complexes and charge transfer in d^{10} complexes can be considered as three members of one family.

3.3.7 s^2 Ions

Ions with outer s^2 configuration are of large importance in the field of luminescence. In principle their spectroscopy is well understood [22]. The influence of the host lattice on these properties is drastic which fact is considerably less understood.

Well-known luminescent ions in this class are Tl^+, Pb^{2+}, Bi^{3+} (all $6s^2$), and Sn^{2+}, Sb^{3+} (both $5s^2$). The influence of the host lattice can be illustrated by comparing $Cs_2NaYCl_6 : Bi^{3+}$ where the emission consists of a narrow band with considerable vibrational structure and a small Stokes shift (800 cm^{-1}) and $LaPO_4 : Bi^{3+}$ where the emission consists of a broad band without any structure at all and with a large Stokes shift (19 200 cm^{-1}). Table 3.3 shows that the Stokes shift varies one order of magnitude which is rather exceptional [23]. Figure 3.20 gives a spectral illustration.

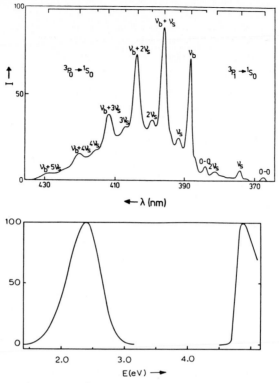

Fig. 3.20. Two different types of Bi^{3+} emission at 4.2 K. Top: $CaO:Bi^{3+}$; halfwidth of emission band about 1200 cm^{-1}; the vibrational structure has been interpreted using a stretching (νs) and a bending (νb) vibration. Bottom: $Bi_2Ge_3O_9$, halfwidth of emission band about 5000 cm^{-1}. The bottom spectrum also gives the excitation spectrum (right hand side), illustrating the enormous Stokes shift ($\sim 20\,000$ cm^{-1})

This large variation has been ascribed to the amount of space available for the Bi^{3+} ion in the lattice. The small Stokes shift case is only observed for Bi^{3+} in six coordination. The Bi^{3+} ion is too large for six coordination and has no possibility to relax to a different equilibrium distance. In a large site the situation is different. It has been proposed that in the ground state the Bi^{3+} goes off center when there is enough space. In this way it obtains its preferred asymmetrical coordination. This is an example of the pseudo Jahn-Teller effect [24]. After optical absorption the ion relaxes to the center of the coordination polyhedron. Due to this large relaxation a large Stokes shift results. Actually there is not much difference between two compounds like $CaWO_4$ and $Bi_4Ge_3O_{12}$ as far as their luminescence is concerned: both show after short-wave ultraviolet excitation an enormous amount of relaxation in the excited state where the bonding conditions have changed considerably in comparison to the ground state.

Evidence for this description of the relaxation of the Bi^{3+} ion stems from EXAFS measurements on $LaPO_4:Bi^{3+}$: in the ground state the Bi^{3+} coordination is more asymmetrical than the La^{3+} coordination. Another piece of evidence is the luminescence of bismuth compounds like $Bi_4Ge_3O_{12}$ where the ground state coordination is known from crystallographic data. Compounds of this type luminesce with large Stokes shifts. The same holds for some lead compounds ($PbAl_2O_4$, $PbGa_2O_4$).

After having presented the wide range of luminescence properties of s^2 ions, we now turn our attention to the excited state. The excited sp configuration yields a lower ^3P state. This state may undergo two different types of interaction, viz. spin-orbit (SO) interaction and Jahn-Teller (JT) (electron-lattice) interaction. The SO interaction will split the ^3P state into the $^3P_0$, $^3P_1$ and $^3P_2$ levels. The strength of this interaction increases with nuclear charge, i.e. from $4s^2$ to $6s^2$ ions and from low to high charge. The JT effect causes also a splitting of the ^3P state due to coupling of this state with vibrational modes. For octahedral coordination these are the ν_2 and ν_5 modes of the MX_6 octahedron ($M = s^2$ ion, $X = $ ligand). If the SO coupling is strong, the influence of the JT effect is expected to decrease.

The importance of the JT effect in the spectroscopy of s^2 ions can be easily observed from the splitting of the $^1S_0 \rightarrow {}^3P_1$, $^1P_1$ absorption transitions [22]. The consequences for the emission are even more drastic. In some cases two emission bands are observed. An example is given in Figure 3.21, where the emission of $YPO_4 : Sb^{3+}$ is given as a function of temperature. The two emissions arise from different minima on the potential energy surface of the relaxed excited state. Figure 3.22 gives a schematic representation. At low temperatures the X minimum which is populated by optical excitation emits. At higher temperatures the barrier can be overcome and emission from the T minimum is also observed. At still higher temperatures, thermal equilibrium between the two minima occurs and the X emission reappears. The occurrence of two different minima depends strongly on the ratio between SO and JT couplings and the site symmetry of the s^2 ion.

Above it was already mentioned that the Te^{4+} ion shows a vibrational progression in its emission spectrum. This has to be ascribed to coupling with the ν_2 mode and indicates a tetragonally distorted excited state. This is also a manifestation of the JT interaction.

In the case of the $6s^2$ ions the SO interaction is so strong that the emission can be interpreted in terms of the SO-split levels $^3P_1$ and $^3P_0$. The vibrational progression in the emission band, if present, is always due to coupling with the symmetric ν_1 mode. At low temperatures the decay time becomes very long (ms range), since the $^3P_0 \rightarrow {}^1S_0$ emission is strongly forbidden. At higher temperatures the $^3P_1$ level becomes thermally occupied and the decay time becomes much faster. This runs parallel with the discussion above on the Eu^{2+} ion (see Fig. 3.16).

Three contributions to the Stokes shift of s^2 ion emission have to be taken into account [25]. This must be done in a multidimensional configurational coordinate diagram and is mentioned here as an illustration of the complicated nature of this emission and, especially, of the relaxation into the excited state. We will restrict ourselves to octahedral coordination.

a. a shift along the Jahn-Teller active modes ν_2 and ν_5. According to the Te^{4+} emission, the shift along ν_2 dominates. This corresponds to a tetragonal distortion.

b. a shift along the symmetrical ν_1 mode. This corresponds to a symmetrical expansion.

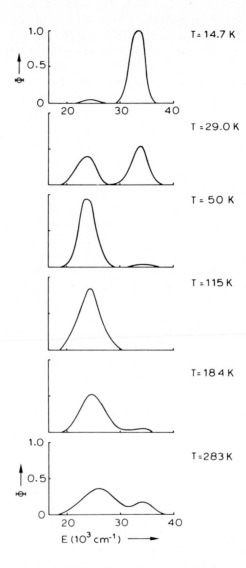

Fig. 3.21. The emission spectrum of $YPO_4 : Sb^{3+}$ as a function of temperature. See also Fig. 3.22. At low temperatures mainly UV emission is observed, at higher temperatures the blue emission dominates, and at room temperature the UV emission reappears. From E.W.J.L. Oomen, thesis, University Utrecht, 1987

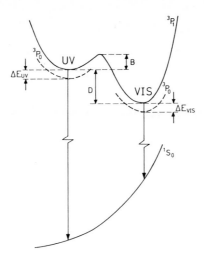

Fig. 3.22. Schematic configurational coordinate diagram for $YPO_4 : Sb^{3+}$. The potential energy curve of the excited state contains two minima. Optical excitation feeds the UV-emitting minimum. At low temperatures UV emission occurs. When the barrier B can be passed thermally, blue emission appears. When thermal back transfer over the barrier $\Delta E + B$ becomes possible, the UV emission reappears

c. a shift along a mode with t_{1u} symmetry which mixes the electronic ground state with excited T_{1u} states (pseudo Jahn-Teller effect). This distorts the ground state, often trigonally.

In the Te^{4+} emission the contribution **a** dominates which is probably true for many $4s^2$ and $5s^2$ ions (JT coupling in the excited state dominating). In the emission of $6s^2$ ions on a small site the contribution **b** dominates. For $6s^2$ ions with a distorted ground state the contribution **c** dominates. However, there are all types of mixtures possible, which makes the s^2 ion emission so complicated.

3.3.8 The U^{6+} ion

Hexavalent uranium played a role in the early experiments by Stokes (1852). This ion can show an intense green luminescence. Originally it was thought that this luminescence could only be obtained from the uranyl (UO_2^{2+}) ion. Later it appeared that octahedral UO_6^{6-} and tetrahedral UO_4^{2-} luminesce as well. The emission color of the latter complex is red. In NaF the introduction of U^{6+} even leads to a whole family of luminescent centers with a general formula $[UO_{6-x}F_x]^{(6-x)-}$ ($x = 0, 1, 2, 3$). These are all octahedral and replace the NaF_6 octahedron in the crystal lattice. Figure 3.23 gives as an example of uranate luminescence the emission spectrum of $Ba_2ZnWO_6 : U^{6+}$.

The U^{6+} ion can formally be considered as a $5f^0$ ion. Indeed the optical transitions involved are of the charge-transfer type. Figure 3.23 shows that we have here the

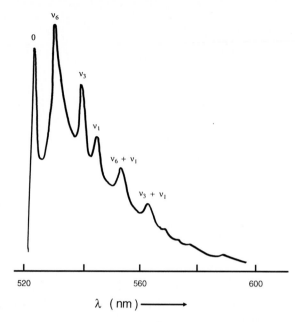

Fig. 3.23. The emission spectrum of $Ba_2ZnWO_6 : U^{6+}$ at 4.2 K. The zero-phonon line is indicated by O, the vibronic lines are indicated by the octahedral vibrational modes with which the coupling with the electronic transition takes place

intermediate-coupling case. An extensive study of the uranyl spectroscopy has been made by Denning [26].

3.3.9 Semiconductors

This book does not aim to treat semiconductors in any detail. However, we have several reasons to mention, at least superficially, emission from semiconductors. These are:

- luminescent semiconductors form an important class of materials with applications in the display field (television, electroluminescence of thin layers, light-emitting diodes)
- compounds made up of the luminescent centers discussed above are sometimes semiconductors, and sometimes not. In this way we meet the interesting boundary regime between semiconductors and insulators.

This section will discuss the following topics: "classic" semiconductors with near-edge and deep-center emission, and subsequently compounds in the boundary regime between semiconductors and insulators. For a more detailed treatment of the first topic, the reader is referred to Ref. [27]; for the second topic we follow a treatment sketched before in Ref. [23].

a. Semiconductors

Semiconductors are characterized by a valence band and a conduction band separated by an energy gap E_g of a few eV. Excitation of the luminescence occurs by exciting electrons to the empty conduction band leaving holes in the completely filled valence band. Emission occurs by electron-hole recombination. However, emission due to recombination of free electrons and holes is exceptional. Usually recombination occurs close to or at defects in the crystal lattice. Phenomenologically it has been the practice to distinguish edge emission, i.e. emission close to the energy E_g, and deep-center emission, i.e. emission at an energy considerably lower than E_g.

Edge emission is due to exciton recombination (Sect. 3.3.1). Usually this emission is due to bound excitons, i.e. an exciton of which either the electron or the hole is trapped at an imperfection in the lattice. The elucidation of the nature of this imperfection is often a difficult task. As an example of such an emission we can mention the exciton emission of GaP:N. Nitrogen is an isoelectronic dopant (on phosphorous sites). The exciton is bound to this nitrogen impurity before it decays. The emission is situated at about 0.02 eV below E_g. Another semiconductor for which exciton emission has been thoroughly studied is CdS.

Another type of recombination in semiconductors is donor-acceptor pair emission. In this type of emission an electron trapped at a donor and a hole trapped at an acceptor recombine. Again GaP is a very nice example (Fig. 3.24). The donor-acceptor pair emission in this figure is due to a $S_P–Zn_{Ga}$ pair transition. The lines are due to the fact that the distance between the donor S_P and the acceptor Zn_{Ga} varies due to their statistical distribution over the lattice, so that the binding energy of electron and hole varies with the distance between the centers where they are trapped.

Also in other semiconductors such donor-acceptor pair emissions have been found. A well-known example is ZnS:Cu,Al where Al_{Zn} is the donor and Cu_{Zn} the acceptor. This material is used as the green-emitting phosphor in color television tubes. The blue emission of ZnS:Al is due to recombination in an associate consisting of a zinc vacancy (acceptor) and Al_{Zn} (donor). Since these centres occur as coupled defects, their distance is restricted to one value only (this is sometimes called a molecular center). Due to strong electron-lattice coupling the emissions from ZnS consist of broad bands.

The emissions of ZnS:Al and ZnS:Cl are practically the same. In both cases we have the same center (although chemically different): an associate of a donor (Al_{Zn} or Cl_S) and an acceptor (V_{Zn}). This illustrates that not the chemical nature of the center determines its luminescence properties, but its position in the forbidden zone. This is very different from the centers discussed in previous sections. Actually reduction of E_g (for example by moving from ZnS to ZnSe) shifts all emissions to a correspondingly lower energy.

Other possibilities for radiative recombination are a free hole that recombines with a trapped electron (Lambe-Klick model) or a free electron that recombines with a trapped hole (Schön-Klasens model). The trapped charge carriers may occupy deep traps, so that the emitted energy is considerably less than E_g. Figure 3.25 reviews the possibilities mentioned here for radiative recombination in a semiconductor.

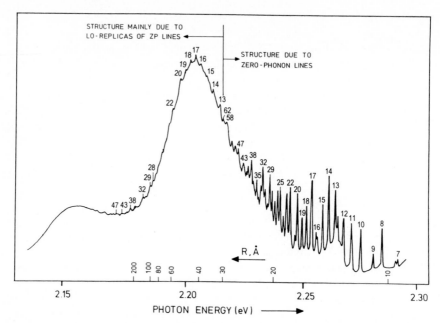

Fig. 3.24. The S_P–Zn_{Ga} donor-acceptor pair emission in GaP at 1.6 K. At the bottom R indicates the donor-acceptor distance of the relevant emitting pair. The lines are indicated by their shell number (shell 1 means nearest-neighbour pairs, etc.). On the right-hand side we see the zero-phonon lines of the individual lines; on the left-hand side we see mainly vibronic lines (in the semiconductor field often called replicas) due to coupling with host-lattice modes. Modified from A.T. Vink, thesis, Technical University Eindhoven, 1974

b. On the Boundary Between Semiconductors and Insulators

The greater part of this chapter has been based on the configurational coordinate model. The luminescent centers were classified according to the value of the Huang-Rhys parameter S (weak, medium or strong coupling). This implies that the interaction between electronic and vibrational transitions was dominant. The system was assumed to have localized electrons.

This neglects another important factor in solids, viz. delocalization. This was in fact only used above when introducing the energy band scheme (in the case of semiconductors). Localization and delocalization are competitive effects as will be shown now.

Consider a system of luminescent centers, each with a two-level energy scheme, with a large distance between the centers. If the distance is decreased, two effects may occur, viz.

– the excited state couples strongly with the vibrations of the system, so that strong relaxation occurs after excitation. This brings the system out of resonance with its neighbours, and promotes localization. This situation is described by the configurational coordinate diagram.

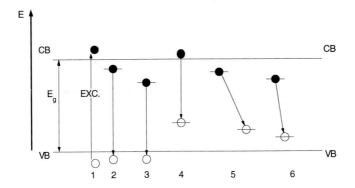

Fig. 3.25. Emission transitions in a semiconductor (schematical representation). The band gap E_g separates the valence band (VB) and the conduction band (CB). Excitation over the band gap (1) creates electrons in CB and holes in VB. Optical recombination is shown in processes 2–6: (2): a free hole recombines with an electron trapped in a shallow trap level (near-edge emission); (3): the same with a deep electron-trapping level; (4): a free electron recombines with a trapped hole; (5): donor-acceptor pair emission; (6): electron-hole recombination in an associate of a donor and an acceptor

– the wave functions of the levels of the individual centers overlap considerably and energy-band formation occurs. This promotes delocalization. The situation is described with the energy band model. Figure 3.26 shows this schematically. The top of the triangle presents the centres at large distance (no interaction). Along the left-hand leg delocalization increases, along the right-hand leg localization. The basis of the triangle presents the solid state with semiconductors with mobile charge carriers on the left-hand side, and insulators on the right-hand side. Along the basis the amount of localization increases from left to right. Simultaneously spectral band width and Stokes shift increases. The semiconductor in this definition is completely pure and should only show free-exciton emission as a sharp line very close to E_g, the band gap energy.

Here are some illustrative examples. The intraconfigurational $4f^n$ transitions of the rare-earth ions are located at the apex of the triangle, because they present, even in a solid, a nearly isolated system.

The compounds $CaWO_4$ and $Bi_4Ge_3O_{12}$ with large relaxation in the excited state and strongly Stokes-shifted emission are on the right-hand side of the base of the triangle (strong electron-lattice coupling, high S value).

Free exciton emission has been observed for $Cs_3Bi_2Br_9$ and TiO_2. They are placed on the left-hand side of the base. This is also true for $CsVO_3$, whereas YVO_4 is on the right-hand side.

The latter example illustrates another difference between compounds on the left- and right-hand side of the basis. The semiconductors (left) have their optical absorption edge at lower energy than the insulators (right). This is because energy band formation reduces the energy difference in the original two-level scheme, whereas

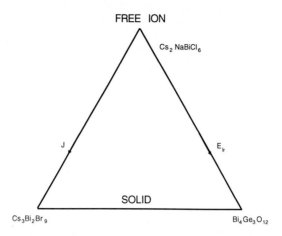

Fig. 3.26. Schematic representation of the transition from the free-ion state (apex) to the condensed-matter state (base). See also text. E_{lr} indicates the electron-lattice coupling, J the orbital overlap. Along the right-hand leg localisation increases, along the left-hand leg delocalisation. The examples concern the Bi^{3+} ion. In $Cs_2NaBiCl_6$ the spectra can be described by a small Huang-Rhys coupling parameter (S), in $Bi_4Ge_3O_{12}$ the value of S is very large, whereas $Cs_3Bi_2Br_9$ is a semiconductor

the parabola offset increases this difference: $CsVO_3$ has its optical absorption edge at ~ 3.4 eV, and YVO_4 at ~ 4 eV. Even more impressive are $Bi_{12}GeO_{20}$ (~ 3 eV) and $Bi_4Ge_3O_{12}$ (~ 5 eV).

Intermediary cases are $SrTiO_3$ and $KTiOPO_4$. Their emissions show small Stokes shifts compared to that of isolated titanate groups. This implies that only at liquid helium temperatures is luminescence observed. The excited state needs only a low thermal activation energy to become mobile.

Interestingly enough, $CsPbCl_3$ is a semiconductor, but $PbCl_2$ an insulator. This is related to the way in which the chloro-lead polyhedra are coupled in the crystal structure. In $CsPbCl_3$ this coupling favours energy band formation.

The triangle of Figure 3.26 has the advantage that it is possible to locate the different luminescent compounds in such a way that their luminescence properties are immediately characterized, albeit approximatively.

3.3.10 Cross-Luminescence

Recently there has been a lot of interest in the luminescence of BaF_2. Its crystals have a potential as a scintillator material (detection of gamma rays, see Chapter 9). They show a luminescence at 220 nm with a very short decay time, viz., 600 ps. This luminescence is of a new type (cross-luminescence). Its nature has been unraveled by Russian investigators [28]. Excitation with about 10 eV excites anion excitons i.e. excitons of which the hole is trapped on F^-. Upon recombination these anion excitons show an emission at about 4.1 eV (300 nm). This is an emission of the type

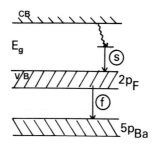

Fig. 3.27. Energy level scheme of BaF$_2$ showing cross luminescence (f). The exciton lumines- cence is indicated (s). See also text

observed for alkali halides (Sect. 3.3.1). Excitation with about 18 eV excites cation excitons. These do not recombine in a simple way, but by a so-called cross-transition: an electron jumps from the F$^-$ ion (2p orbital) into the hole in the 5p orbital of Ba^{2+} (see Figure 3.27). This is accompanied by emission at about 5.7 eV (220 nm), and weaker emissions at even higher energy. Since the energy difference between the 2p (F$^-$) and the 5p (Ba^{2+}) energy band is less than the band gap ($\sim$ 10 eV), the corresponding emission is observed as part of the intrinsic emission of BaF$_2$. The 200 nm emission shows practically no temperature quenching up to room temperature, whereas the 300 nm emission is for the greater part quenched under these conditions.

Other compounds for which this phenomenon has been found are CsCl and CsBr, and KF, KMgF$_3$, KCaF$_3$, and K$_2$YF$_5$.

3.4 Afterglow

Afterglow is the phenomenon that luminescence can still be observed a long time after the end of the excitation pulse. A long time here is defined as a time much longer than the decay time τ_R of the luminescence (Sect. 3.2). Everybody knows this phenomenon from the luminescent lamp that still glows after being switched off. For certain applications the level of afterglow should be negligbly low.

Afterglow is due to the phenomenon that radiant recombination of electrons and holes is sometimes considerably delayed due to trapping of electrons or holes. Fig- ure 3.28 gives a simple illustration. A semiconductor contains, next to the luminescent centers, also centers which trap electrons. Excitation with energy above E$_g$ yields free electrons and holes. Let us assume that the holes are trapped by the luminescent cen- ter, whereas the electrons in the conduction band recombine with the holes yielding emission.

However, part of the electrons are trapped in the electron trap center, from where they escape thermally after some time. Only then they recombine with a trapped hole. The corresponding emission occurs with considerable delay and is called afterglow.

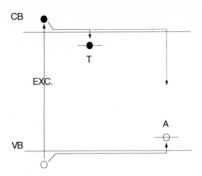

Fig. 3.28. The afterglow mechanism. Excitation (EXC) creates free electrons and holes. The holes are trapped at the level A where recombination with an electron yields radiative recombination (emission). However, the electron is trapped at the level T, so that it arrives only at A after considerable delay. Therefore afterglow results

A simple example is $Y_3Ga_5O_{12}:Cr^{3+}$. Holes are trapped by Cr^{3+}, forming Cr^{4+}. Electrons recombine with Cr^{4+}, yielding excited Cr^{3+} ions which emit. However, part of the electrons are trapped at oxygen vacancies from where they escape later to yield afterglow [29].

3.5 Thermoluminescence

If the system depicted in Figure 3.28 is irradiated with $E > E_g$ at temperatures which are too low to allow electrons to escape from the traps, the traps are filled and can only be emptied by increasing the temperature. If the temperature is now increased regularly, emission from the luminescent center will occur at temperatures where the traps are emptied thermally. The resulting curve of luminescence intensity versus temperature is called a glow curve. An example is given in Figure 3.29. From the position and shape of the glow peaks, information can be obtained about the traps [30]. Note that this is thermal stimulation of luminescence (and not thermal excitation, since the excitation has occurred before, viz. during filling of the traps).

It is also possible to perform optical stimulation, viz. by irradiation with light the energy of which is sufficient to excite the electron from the trap into the conduction band. In certain applications this photostimulation plays an important role (see Chapter 8).

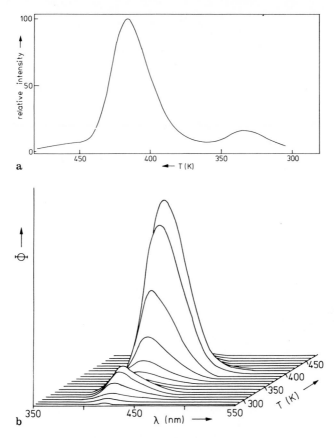

Fig. 3.29. Glow curve of X-ray irradiated $Ba_2B_5O_9Br:Eu^{2+}$ (*top*). The luminescence intensity is given as a function of temperature. The presence of two glow peaks indicates the presence of two traps with different trap depths in the host lattice. The lower picture presents the thermo-luminescence emission spectra of the same sample. Actually the top curve is a cross section of the bottom picture. For reference, see figure 3.15.

3.6. Stimulated Emission

It should be realized that absorption and emission as discussed in Chapters 2 and 3 are different processes, since the former needs a radiation field and the latter not. Einstein considered the problem of transition rates in the presence of a radiation field [31]. For the transition rate from lower to upper level he wrote $w = B\rho$, where B is the Einstein coefficient of (stimulated) absorption and ρ is the radiation density.

The radiation field induces also a transition from the upper to the lower state (stimulated emission). The rate is $w' = B'\rho$, where B' is the Einstein coefficient of stimulated emission. The rate of spontaneous emission is $w'' = A$, with A the coefficient of spontaneous emission (note the absence of ρ in this expression). This is

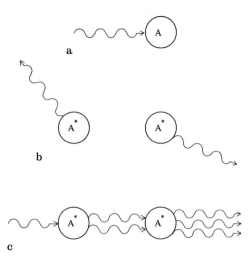

Fig. 3.30. Schematic representation of *a* absorption (A absorbs radiation and reaches the excited state A* which is not shown in *a*); *b* spontaneous emission from two species A* (the two emissions are not correlated); *c* stimulated emission (incoming radiation on the left-hand side "forces" two A* species to return to the ground state; the outcoming radiation on the right-hand side is amplified by a factor 3 as far as the amplitude is concerned)

shown schematically in Figure 3.30. It can be derived that $B' = B$ and $A = 8\pi h\nu^3 c^{-3}B$ (ν is the frequency of the emitted radiation, c is the speed of light).

If the two levels have an energy separation of, for example, 20 000 cm^{-1} (visible light), the lower level will be predominantly occupied, i.e. stimulated emission can be neglected. This justifies the approach of Chapters 2 and 3. In the case of three or more levels, the situation in which one of the higher levels has a higher occupation than the ground level (population inversion) can, under special conditions, be realized.

This can be illustrated on the Cr^{3+} ion in a strong crystal field. Its energy level scheme is given in Figure 3.31. Upon irradiation into the $^4A_2 \rightarrow {}^4T_2$ transition with high intensity, population inversion between the ^2E and $^4A_2$ levels can be obtained. This is due to the fact that the $^4T_2$ level empties rapidly into the ^2E level which in turn has a very long life time (ms) in view of the spin selection rule.

If one of the excited Cr^{3+} ions decays spontaneously, the emitted photon will stimulate other excited Cr^{3+} ions to decay by stimulated emission, so that there is an amplification of the original photon. This is the principle of laser action (laser stands for light amplification by stimulated emission of radiation). The amplification ends when the population inversion is over, since absorption then dominates stimulated emission.

Therefore, a four-level laser offers an important advantage, viz. population inversion is easier to obtain and to maintain (see Figure 3.32): level 4 should decay rapidly to the ground level 1, so that any population of level 3 corresponds to population inversion. The Nd^{3+} ion serves in this way.

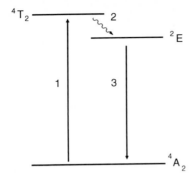

Fig. 3.31. The ruby three-level laser. The pumping transition is 1, the lasing transition 3. The nonradiative transition 2 is fast relative to the radiative inverse of 1. The level notations are for Cr^{3+}. The orders of magnitude of the relevant rates are $p(^4T_2 \rightarrow {}^4A_2) = 10^5 \text{ s}^{-1}$, $p(^4T_2 \rightarrow {}^2E) = 10^7 \text{ s}^{-1}$, $p(^2E \rightarrow {}^4A_2) = 10^2 \text{ s}^{-1}$

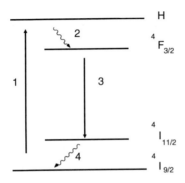

Fig. 3.32. The four-level laser scheme. Pumping transition is 1, lasing transition 3, nonradiative transitions 2 and 4. On the right-hand side the level notation for Nd^{3+} for the case of the 1064 nm laser action is given. H denotes levels above $^4F_{3/2}$; further $^4I_{11/2}$ cannot be thermally populated since it is about 2000 cm^{-1} above the ground state

This book does not intend to deal with laser physics and all the possible types of lasers. Two remarks are still worth noting:

- By using broad-band emission as depicted in Figure 3.1, it is possible to make a tunable laser. This is a four-level laser (level 1 is the lowest vibrational level of the ground state parabola, level 2 are the vibrational levels of the excited state reached after optical absorption, level 3 is the lowest vibrational level of the excited state parabola, and level 4 is one of the vibrational levels of the ground state parabola reached after emission). Tunability is achieved by selecting level 4.
- A laser material should satisfy several requirements. One of these is that it shows efficient luminescence. Further, laser materials should be characterized by careful spectroscopic measurements.

In view of the latter remark the phenomenon of stimulated emission has been mentioned, and certain laser materials will be mentioned later in this book. History illustrates this by the fact that ruby, investigated long ago by Becquerel, is the material on which the first solid state laser was based. In this chapter the return from the excited state to the ground state was assumed to be radiative only. The next chapter considers nonradiative transitions.

References

1. Henderson B, Imbusch GF (1989) Optical spectroscopy of inorganic solids. Clarendon, Oxford
2. Donker H, Smit WMA, Blasse G (1989) J. Phys. Chem. Solids 50:603; Wernicke R, Kupka H, Ensslin W, Schmidtke HH (1980) Chem. Phys. 47:235
3. Wolfert A, Oomen EWJL, Blasse G (1985) J. Solid State Chem. 59:280
4. Atkins PW (1990) Physical chemistry, 4th ed. Oxford University Press, Oxford
5. Tanimura K, Makimura T, Shibata T, Itoh N, Tokizaki T, Iwai S, Nakamura A (1993) Proc. int. conf. defects insulating materials. S. Nordkirchen, World Scientific, Singapore, p 84; Puchin VE, Shluger AL, Tanimura K, Stok N (1993) Phys Rev B47:6226
6. Blasse G (1992) Int Rev Phys Chem 11:71
7. Brixner LH, Blasse G (1989) Chem. Phys. Letters 157:283
8. Judd BR (1962) Phys. Rev. 127:750; Ofelt GS (1962) J. Chem. Phys. 37:511
9. Mello Donega C de, Meijerink A, Blasse G (1992) J. Phys.: Cond. Matter 4:8889
10. Di Bartolo B (1968) Optical interactions in solids, Wiley, New York
11. Schotanus P, van Eijk CWE, Hollander RW (1988) Nucl. Instr. Methods A 272:913; Schotanus P, Dorenbos P, van Eijk CWE, Hollander RW (1989) Nucl. Instr. Methods A 284:531
12. Meijerink A, Nuyten J, Blasse G (1989) J. Luminescence 44:19
13. Blasse G, Dirksen GJ, Meijerink A (1990) Chem. Phys. Letters 167:41
14. Blasse G, de Korte PHM (1981) J. Inorg. Nucl. Chem. 43:1505
15. Herren M, Güdel HU, Albrecht C, Reinen D (1991) Chem. Phys. Letters 183:98
16. Blasse G (1980) Structure and Bonding 42:1
17. Barendswaard W, van der Waals JH (1986) Molec. Phys. 59:337; Barendswaard W, van Tol J, Weber RT, van der Waals JH (1989) Molec. Phys. 67:651
18. Hazenkamp MF, thesis, University Utrecht, 1992
19. Blasse G (1991) Structure and Bonding 76:153
20. Tol van J, van der Waals JH (1992) Chem. Phys. Letters 194:288
21. Bruin de TJM, Wiegel M, Dirksen GJ, Blasse G (1993) J. Solid State Chem. 107:397
22. Ranfagni A, Mugni M, Bacci M, Viliani G, Fontana MP (1983) Adv. Physics 32:823
23. Blasse G (1988) Progress Solid State Chem. 18:79
24. Bersuker IB, Polinger VZ (1989) Vibronic interactions in molecules and crystals, Springer, Berlin
25. Blasse G, Topics in Current Chemistry, in press.
26. Denning RG (1992) Structure and Bonding 79:215
27. Kitai AH (ed) Solid state luminescence. Theory Materials and Devices, Chapman and Hall, 1993
28. Valbis YaA, Rachko ZA, Yansons YaL (1985) JETP Letters 42:172; Aleksandrov YuM, Makhov VN, Rodnyl PA, Syreinshchikova TI, Yakimenko MN (1984) Sov. Phys. Solid State 26: 1734
29. Grabmaier BC (1993) Proc. int. conf. defects insulating materials, Nordkirchen, World Scientific, Singapore, p 350; Blasse G, Grabmaier BC, Ostertag M (1993) J. Alloys Compounds 200: 17
30. McKeever SWS (1985) Thermoluminescence of solids, Cambridge University, Cambridge
31. For a simple account, see ref. [4]. For a more detailed account ref. [1]

CHAPTER 4

Nonradiative Transitions

4.1 Introduction

Radiative return from the excited state to the ground state (Chapter 3) is not the only possibility of completing the cycle. The alternative is nonradiative return, i.e. a return without emission of radiation. Nonradiative processes will always compete with radiative processes. Since one of the most important requirements for a luminescent material is a high light output, it is imperative that in such a material the radiative processes have a much higher probability than the nonradiative ones.

All energy absorbed by the material which is not emitted as radiation (luminescence) is dissipated to the crystal lattice (radiationless processes). It is, therefore, imperative to suppress those radiationless processes which compete with the emission process. There are, however, also nonradiative processes which favour a high light output, viz. those which ensure a more effective feeding of the luminescent activator and/or population of the emitting level.

Table 4.1 gives the quantum efficiencies (see Sect. 4.3) of some important photoluminescent materials. Note that the maximum obtainable value of 100% is not obtained by far. The complicated composition $NaGdF_4$–Ce,Tb gives the nearest approach to this value. As we will see below, this is due to the fact that the radiationless feeding-processes of the emitting Tb^{3+} ion are, although complicated, very effective, and the radiationless processes competing with the emission very ineffective. In $CaWO_4$, on the other hand, the emitting tungstate group is excited directly, and the 70% quantum efficiency is due to a ratio of the radiative and nonradiative decay rates of the emitting level of about 2:1.

Table 4.1. Quantum efficiencies q of some photoluminescent materials under ultraviolet excitation at room temperature

Material	q(%)
Zn_2SiO_4–Mn^{2+}	70
YVO_4–Eu^{3+}	70
$CaWO_4$	70
$Ca_5(PO_4)_3(F,Cl)$–Sb^{3+},Mn^{2+}	71
$NaGdF_4$–Ce^{3+}, Tb^{3+}	95

These observations illustrate that it is important to have a good understanding of radiationless processes in order to understand, to improve, and to predict luminescent materials. There are several approaches to this difficult problem [1,2]. In our opinion all of these are useful. One of these is a completely theoretical approach. This may nowadays be of great use in understanding specific, simple situations. However, it cannot be used with much success to account for the properties of a large number of different materials.

There are also general, approximative methods for studying nonradiative processes. A good example is the book by Struck and Fonger [3]. Our approach is probably the simplest one. The main characteristic is that the physical parameters are kept as simple as possible, whereas the chemical parameters are widely varied, i.e. by studying a large number of different materials we try to find out in how far the simple physical model is still able to account for the results. The great advantages of this method are its simplicity and predictive ability.

This chapter is organized as follows. In Sect. 4.2 we consider one isolated luminescent center and the possible nonradiative transitions in such a center. In Sect. 4.3 we define the several ways to express the efficiency of a luminescent material. In Sect. 4.4 we discuss the conversion efficiency of a luminescent material upon host lattice excitation over the energy band gap E_g. Finally Sections 4.5 and 4.6 show some other possiblities for nonradiative transitions.

4.2 Nonradiative Transitions in an Isolated Luminescent Centre

In Chapter 3 it was assumed that the return from the excited state to the ground state is radiative. This is usually not the case. Actually there are many centers which do not luminesce at all. Let us consider the configurational coordinate diagrams of Figure 4.1 in order to understand the relevant physical processes. Figure 4.1a presents essentially the same as Figure 3.1. Absorption and emission transitions are possible and Stokes shifted relative to each other. The relaxed-excited-state may, however, reach the crossing of the two parabolas if the temperature is high enough. Via the crossing it is possible to return to the ground state in a nonradiative manner (see arrow in Figure 4.1a). The excitation energy is then completely given up as heat to the lattice. This model accounts for the thermal quenching of luminescence (the latter being essentially a low-temperature phenomenon). The transition from parabola e to parabola g will not be considered further here. Essentially it is a transition between two (nearly) resonant vibrational levels, one belonging to e, the other to g. It will be clear that the larger the offset between the parabolas, the easier the transition occurs.

In Figure 4.1b, the parabolas are parallel (S = 0) and will never cross. It is impossible to reach the ground state in the way described for Figure 4.1a. However, also here nonradiative return to the ground state is possible if certain conditions are fulfilled, viz. the energy difference ΔE is equal to or less than 4–5 times the higher

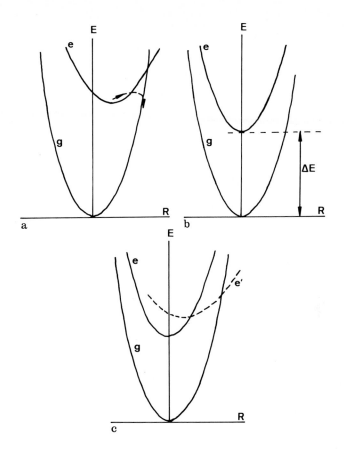

Fig. 4.1. Configurational coordinate diagrams illustrating nonradiative transitions. The ground state parabola is indicated by g, the excited state parabolas by e and e'. See also text. In (a), the arrow indicates a nonradiative transition from e to g, which quenches the luminescence at higher temperatures. In (b), ΔE is the energy difference between e and g. In (c), excitation is into e'; this level feeds the emitting level e

vibrational frequency of the surroundings. In that case, this amount of energy can simultaneously excite a few high-energy vibrations, and is then lost for the radiative process. Usually this nonradiative process is called multi-phonon emission.

In Figure 4.1c both processes are possible in a three-parabolas-diagram. The parallel parabolas will belong to the same configuration, so that they are connected by forbidden optical transitions only. The third one originates from a different configuration and is probably connected to the ground state by an allowed transition. This situation occurs often. Excitation (absorption) occurs now from the ground state to the highest parabola in the allowed transition. From here the system relaxes to the relaxed excited state of the second parabola. Figure 4.1c shows that the nonradiative transition between the two upper parabolas is easy. Emission occurs now from the second parabola (line emission). This situation is found for $Al_2O_3 : Cr^{3+}$ ($^4A_2 \rightarrow {}^4T_2$ excitation, $^4T_2 \rightarrow {}^2E$ relaxation, $^2E \rightarrow {}^4A_2$ emission), Eu^{3+} ($^7F \rightarrow$ charge-transfer-state excitation, charge-transfer-state to ^5D relaxation, $^5D \rightarrow {}^7F$ emission), and Tb^{3+} ($^7F \rightarrow 4f^75d$ excitation, $4f^75d \rightarrow {}^5D$ relaxation, $^5D \rightarrow {}^7F$ emission).

In general the temperature dependence of the nonradiative processes is reasonably well understood. However, the magnitude of the nonradiative rate is not, and can also not be calculated with any accuracy except for the weak-coupling case. The reason for this is that the temperature dependence stems from the phonon statistics which is known. However, the physical processes are not accurately known. Especially the deviation from parabolic behaviour in the configurational coordinate diagram (anharmonicity) may influence the nonradiative rate with many powers of ten. However, it will be clear that the offset between the two parabolas (ΔR) is a very important parameter for the nonradiative transition rate. This rate will increase dramatically if ΔR becomes larger.

We consider first the weak-coupling case ($S \sim 0$), and subsequently the intermediate- and strong-coupling cases ($S \gg 0$).

4.2.1 The Weak-Coupling Case

Nonradiative transitions in the weak-coupling approximation are probably the best understood nonradiative processes. The experimental data relate mainly to the rare-earth ions, as far as their sharp-line transitions are considered (i.e. intra-$4f^n$-configuration transitions). The topic has been discussed in books and review papers [1,2,4,5]. We summarize as follows:

For transitions between levels of a $4f^n$ configuration the temperature dependence of the nonradiative rate is given by

$$W(T) = W(0)(n+1)^p \tag{4.1}$$

where $W(T)$ is the rate at temperature T, $p = \frac{\Delta E}{h\nu}$, ΔE the energy difference between the levels involved, and

$$n = [\exp(\frac{h\nu}{kT}) - 1]^{-1}. \tag{4.2}$$

$W(0)$ is large for low p, i.e. for small ΔE or high vibrational frequencies. Further

$$W(0) = \beta \exp[-(\Delta E - 2h\nu_{max})\alpha], \tag{4.3}$$

with α and β constants, and ν_{max} the highest available vibrational frequency of the surroundings of the rare earth ion. This is the energy-gap law in the revised form by Van Dijk and Schuurmans [6], which makes it possible to calculate W with an accuracy of one order of magnitude.

Let us illustrate this with some consequences. In aqueous solutions or in hydrates the rare earth ions do not emit efficiently with the exception of Gd^{3+} ($\Delta E = 32\,000$ cm^{-1}, $\nu_{max} \simeq 3500$ cm^{-1}). For Tb^{3+} ($\Delta E \simeq 15|000$ cm^{-1}), and especially

Eu^{3+} ($\Delta E \simeq 12|000$ cm^{-1}), the quantum efficiencies (q) are depressed, and the other rare earth ions do practically not emit at all. For solids this can be nicely studied in the host lattice $NaLa(SO_4)_2 \cdot H_2O$ where the rare earth site is coordinated to one H_2O molecule only. The q values are as follows: Gd^{3+} q = 100%, Tb^{3+} 70%, Eu^{3+} 10%, $Sm^{3+} \sim 1\%$, $Dy^{3+} \sim 1\%$.

In the classic oxide glasses the rare-earth ions do not emit efficiently, since $\nu_{max} \simeq 1000\text{–}1200$ cm^{-1} (silicates, borates, phosphates). Only Gd^{3+}, Tb^{3+} and Eu^{3+} show efficient luminescence. This situation changes drastically by using fluoride or chalcogenide glasses, where ν_{max} is considerably lower. A very interesting lattice in this connection is $Eu_2Mg_3(NO_3)_{12} \cdot 24H_2O$. At first sight the large number of water molecules is expected to quench the Eu^{3+} emission completely. However, the Eu^{3+} ions are bidentately coordinated by six nitrate ions which shield them from the water molecules. The quantum efficiency is high.

Ions like Eu^{3+} and Tb^{3+} may emit from higher excited states: Eu^{3+} not only from $^5D_0$ (red), but also from $^5D_1$ (green) and $^5D_2$ (blue). However, this depends critically upon the host lattice. In Y_2O_3–Eu^{3+}, for example, all these emissions are observed, since $\nu_{max} \simeq 600$ cm^{-1}. In borates and silicates, however, they are not.

This can be well studied by laser spectroscopy. An example is $NaGdTiO_4 : Eu^{3+}$ [7]. The time dependence of the Eu^{3+} emission in $NaGdTiO_4$ upon excitation into the $^5D_1$ level of Eu^{3+} is as follows: 10 μs after the excitation pulse the emission originates mainly from the $^5D_1$ level, but after longer times the $^5D_1$ intensity has decreased and that of $^5D_0$ increased (see Figure 4.2). The decay curves of the $^5D_0$ emission show a build-up. From these data the rate for $^5D_1 \rightarrow {}^5D_0$ decay is found to be $1.3 \times 10^4 s^{-1}$ at 4.2 K. Its temperature dependence is given by $(n + 1)^p$ as argued above. The value of p turns out to be 5, the vibrational frequency involved being 347 cm^{-1}. This corresponds to the Eu–O stretching vibration. At 300 K this nonradiative rate is about $4 \times 10^4 s^{-1}$. The values of the nonradiative rate exceed that of the radiative $^5D_1$–^7F_J rate ($\sim 10^3 s^{-1}$), so that the nonradiative process dominates and the emission occurs mainly from the $^5D_0$ level. In compounds with higher phonon frequencies available (e.g. borates, silicates), the $^5D_1$ emission of Eu^{3+} is usually hardly detectable, since the value of p is then much lower.

The Tb^{3+} ion may not only emit from $^5D_4$ (green), but also from $^5D_3$ (blue). ΔE is about 5000 cm^{-1}, much larger than in the case of Eu^{3+}. Diluted Tb^{3+} systems show, therefore, always some blue Tb^{3+} emission, unless ν_{max} is very high. Please note that in these examples we consider only ions interacting with their immediate surroundings and not with other luminescent ions. In Chapter 5 the quenching of higher level emission due to interaction with another centre of the same kind will be discussed (cross relaxation).

Finally we pay attention to Gd^{3+} ($4f^7$). Its energy level scheme is given in Figure 4.3. The excited levels are in the ultraviolet and the corresponding transitions have low oscillator strengths. As a consequence, accurate spectroscopy can only be performed with ultraviolet tunable lasers and/or X-ray excitation.

The emission transition $^6P_J \rightarrow {}^8S$ occurs over an energy gap which is about 32 000 cm^{-1}. Nonradiative transitions cannot compete with this radiative one, because it occurs over such a large ΔE. Even water molecules ($\nu \sim 3500$ cm^{-1}) are not able

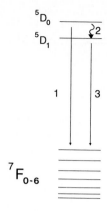

Fig. 4.2. Energy level scheme of Eu^{3+}. The $^5D_1$ level can decay radiatively to the ^7F_J levels (arrow 1), or nonradiatively to the $^5D_0$ level (arrow 2). The latter decay is followed by $^5D_0 \rightarrow$ ^7F_J emission (arrow 3). The amount of $^5D_1$ emission is determined by the rate ratio of processes 1 and 2

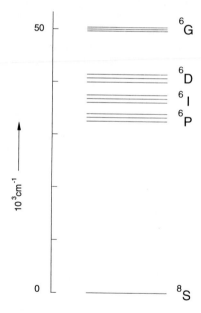

Fig. 4.3. Energy level scheme of the Gd^{3+} ion

to quench the Gd^{3+} emission. The emission can only be quenched by transfer to other luminescent centres (see Chapter 5).

In some host lattices emission has also been observed from the higher excited levels ^6I_J, ^6D_J, and even ^6G_J. However, this depends strongly on the host lattice. If

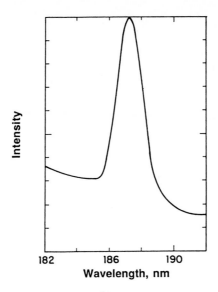

Fig. 4.4. The 187 nm emission line of Gd^{3+} (in Y_2O_3). Excitation is by X rays

the maximum vibrational frequency is low, and the host lattice transparant, emission can be observed from all these levels, up to the ^6G_J level (at about 205 nm) (see Fig. 3.9). There is even evidence for a higher level with $\sim$ 185 nm emission. Figure 4.4 gives an example. In the case of borates and hydrates, however, all these emissions are quenched in favour of the ^6P_J emission. This is a clear demonstration that higher-frequency vibrations promote the radiationless transitions to the ^6P_J levels.

4.2.2 The Intermediate- and Strong-Coupling Cases

This section starts by illustrating how important the value of the parabola offset (ΔR) is for the nonradiative transition rate. We will use several examples. First we will consider $CaWO_4$, a well-known X-ray phosphor for more than seventy-five years. The luminescent group is the tungstate group, a distinct example of a center for which the strong-coupling scheme holds (see Chapters 2 and 3). $CaWO_4$ is a very efficient luminescent material at room temperature. The isomorphous $SrWO_4$, however, does not emit at that temperature, but has to be cooled down in order to reach a reasonably efficient luminescence. Also $BaWO_4$ has the same crystal structure, but even at 4.2 K it does not emit with high efficiency. Nevertheless the ground state properties of the tungstate group in these compounds (distances, vibrational frequencies) are practically equal. The strongly different radiationless processes have to be ascribed to a difference in ΔR, i.e. the parabola offset. Since the ionic radii of Ca^{2+}, Sr^{2+}, Ba^{2+} increase in that order, it seems obvious to assume that this is the reason why the offset increases, i.e. why the rate of the radiationless processes increases, as observed experimentally. The softer the surroundings, the larger is ΔR. Here it is assumed that the presence of ions

Table 4.2. Thermal quenching of the uranate luminescence of ordered perovskites A_2BWO_6–U^{6+}. Compare also Figure 4.5. Data from Ref. [8]

A_2BWO_6–U		$T_q(K)$[1]	$\Delta R(a.u.)$[2]
A=Ba	B=Ba	180	10.9
Ba	Sr	240	10.6
Ba	Ca	310	10.2
Ba	Mg	350	10.0
Sr	Mg	350	10.0
Ca	Mg	350	10.0

1. Quenching temperature of the uranate luminescence.
2. ΔR in arbitrary units, calculated by the Struck and Fonger method [3]. ΔR is arbitrarily put at 10.0 for Ba_2MgWO_6.

with a large radius around the luminescent center is equivalent to soft surroundings. This seems a reasonable assumption.

There is a more impressive experiment to prove this simple model, viz. the luminescence in the ordered perovskites A_2BWO_6 where A and B are alkaline earth ions. Table 4.2 presents the quenching temperatures of the luminescence of the UO_6 group in these lattices [8]. Those for the WO_6 group run parallel. These temperatures are used as a measure of the radiationless processes. The table shows that the radiationless rate does not depend on the nature of the ion A, whereas that of the ion B determines the value of this rate: the smaller the B ion, the higher the quenching temperature.

Figure 4.5 shows that an expansion of the luminescent UO_6 (or WO_6) octahedron (i.e. the parabola offset) is not directly counteracted by the A ion. However, the B ions are immediately involved, the angle U(W)–O–B being 180°. Table 4.2 also shows relative values of ΔR calculated from the Struck and Fonger model [3]. These scale indeed according to prediction. Note that the total change in ΔR is less than 10%. Since ΔR is less than 0.1 Å for the uranate group, the variation in ΔR in this series of compounds is less than 0.01 Å. This shows that small changes in ΔR result in drastic changes in the nonradiative rates.

It is well known that luminescent materials with high quantum efficiencies and quenching temperatures usually have stiff lattices, so that expansion upon excitation is counteracted, i.e. ΔR is as small as possible.

Table 4.3 shows for a series of borates how the Stokes shift, i.e. ΔR, increases if the size of the host lattice cation increases [9]. In $ScBO_3$ the rare earth ions are strongly compressed and the surroundings are stiff. Small Stokes shifts result for Ce^{3+}, Pr^{3+} and Bi^{3+}, but not for the much smaller Sb^{3+}. Note also, that the Stokes shift of the $4f$–$5d$ transitions is less sensitive to the surroundings than that of the $5s$–$5p$ transitions.

Part of solid state chemistry is nowadays involved with what is called soft chemistry or soft materials. As a matter of fact these are not expected to luminesce, at least not when the luminescent centers are broad-band emitters. This has been shown to be the case, for example, for the isomorphous $Al_2(WO_4)_3$, $Sc_2(WO_4)_3$ and $Zr_2(PO_4)_2SO_4$. The Stokes shift of the tungstate and zirconate luminescence in these materials is enor-

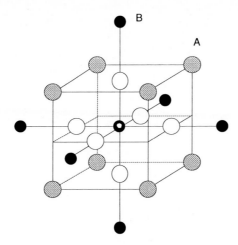

Fig. 4.5. The surroundings of the luminescent UO_6 or WO_6 group in the ordered perovskites A_2BWO_6. The luminescent center consists of the central black, white-centered ion and the six open ions (oxygen). This center is surrounded by a cube of hatched A ions and an octahedron of black B ions

Table 4.3. Stokes shift (10^3 cm^{-1}) of the band emission of some trivalent ions in the orthoborates MBO_3 (M=Sc,Y,La) After Ref. [9]

	ScBO$_3$	YBO$_3$*	LaBO$_3$
Ce^{3+}(4f^1)	1.2	2.0	2.4
Pr^{3+}(4f^2)	1.5	1.8	3.0
Sb^{3+}(5s^2)	7.9	$\left(\begin{matrix}14.5\\16.0\end{matrix}\right.$	19.5
Bi^{3+}(6s^2)	1.8	$\left.\begin{matrix}5.1\\7.7\end{matrix}\right)$	9.3

* This lattice contains two sites for Y.

mous, viz. some $20\,000$ cm^{-1}. The quantum efficiencies, even at 4.2 K are low. This is all due to a large value of ΔR due to the soft surroundings of the luminescent centre.

This model in which the nonradiative transitions can be suppressed by stiff surroundings, can be most elegantly tested by studying the luminescence of rare-earth cryptates [10].

The cryptand ligands are organic cages. They were synthesized for the first time by Lehn who obtained the Nobel prize for Chemistry in 1987 for this achievement (together with Cram and Pederson). Figure 4.6 gives two examples. The 2.2.1 cryptand is just large enough to contain the Ce^{3+} ion, i.e. upon excitation the Ce^{3+} ion has not much space to expand. In fact the [Ce $\subset$ 2.2.1]$^{3+}$ cryptate shows an efficient (broad-band) emission at room temperature with a small Stokes shift, in the solid state as well as in aqueous solution. The [Ce $\subset$ 2.2.2]$^{3+}$ cryptate luminescence has a much larger Stokes shift. As a matter of fact the 2.2.2 cryptand offers a larger hole than the 2.2.1 cryptand.

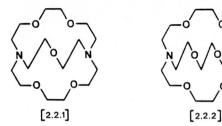

[2.2.1]   [2.2.2]

Fig. 4.6. The [2.2.1] and the [2.2.2] cryptands. The empty corners contain a–CH$_2$ group. These molecules have the shape of a cage in which a trivalent rare earth ion can be introduced

Table 4.4. The Stokes shift of the Ce^{3+} emission for several luminescent compositions/species

Composition/species	Stokes shift (cm^{-1})
[Ce^{3+} $\subset$ 2.2.1]	2100
[Ce^{3+} $\subset$ 2.2.2]	4000
Ce^{3+} in aqueous solution	5000
Y$_3$Al$_5$O$_{12}$–Ce^{3+}	3800
Y$_2$SiO$_5$–Ce^{3+}	2500
ScBO$_3$–Ce^{3+}	1200

Table 4.4 shows the Stokes shift of the Ce^{3+} emission in several surroundings. In the 2.2.1 cryptand the Ce^{3+} Stokes shift is smaller than in some commercial Ce^{3+}-activated phosphors (Y$_2$SiO$_5$–Ce, Ca$_2$AlSiO$_7$–Ce). It becomes very small in ScBO$_3$ (see above) and in CaF$_2$ and CaSO$_4$ where the Ce^{3+} ion carries an effectively positive charge which will make the Ca site smaller than it is on basis of the Ca^{2+} ionic radius.

The experiments treated above give ample evidence that ΔR should be as small as possible if an efficient luminescent material is required. Not only the value of ΔR is of importance. This can be shown by a simple model calculation [11]. We consider a luminescent material in which the luminescent centre is described with a single-configurational coordinate diagram with two parabolas with equal force constants k. The offset is ΔR, the vibrational frequency hν, and the energy difference between the parabolas E$_{zp}$. By introducing certain values for other parameters involved and following the Struck and Fonger approach [3], the temperature dependence of the luminescence efficiency can be calculated (see Fig. 4.7).

Note that the larger ΔR, the lower the quenching temperature of the luminescence, i.e. the more important the radiationless processes. High values of hν also promote radiationless decay. For h$\nu = 600$ cm^{-1} the low-temperature value of the efficiency remains far below the maximum possible value, i.e. even at 0K the rate of the nonradiative processes is about equal to that of the radiative ones. This is due to tunnelling from the excited state parabola to the ground state parabola (Fig. 4.8). A decreasing value of E$_{zp}$ also yields a lower quenching temperature.

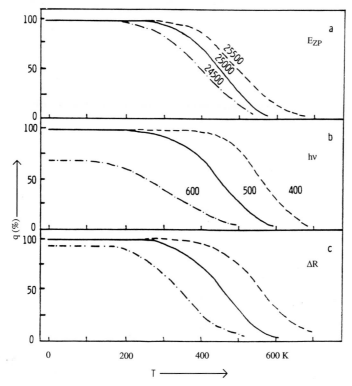

Fig. 4.7. Temperature dependence of the luminescence quantum efficiency (q) according to calculations on a model phosphor system. In (*a*) the energy difference between the parabola minima (E_{zp}) is varied (values in cm^{-1}). In (*b*) the vibrational frequency is varied (values in cm^{-1}). In (*c*) the parabola offset (ΔR) varies; it decreases from left to right by 6%

  The latter is experimentally confirmed by a comparison of $CaWO_4$ and $CaMoO_4$. These compounds are comparable in many aspects. A striking difference is that the energy levels of the tungstate group lie some 5000 cm^{-1} higher than those of the molybdate group. For the room-temperature luminescence of these compounds such a difference has an important consequence: whereas $CaWO_4$ shows an efficient blue emission, the greenish emission of $CaMoO_4$ is partly quenched. This observation can be generalized: broad band emission is the more efficient, the shorter the maximum wavelengths of the excitation and emission bands are.

  Let us now consider luminescent centers with three-parabola diagrams (Fig. 4.1.c). A clear example is the charge-transfer excitation of the Eu^{3+} luminescence, a process which is of large importance for applications. Consider the red phosphor in the three-color luminescent lamps. Its composition is $Y_2O_3 : Eu$. Excitation at 254 nm in the charge- transfer state is followed by efficient red emission ($^5D_0-^7F_2$) within the $_4f^6$ configuration. Figure 4.9 shows the relevant configurational coordinate diagram.

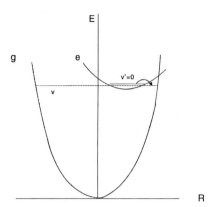

Fig. 4.8. The system can tunnel (arrow) from the lowest vibrational level $v' = 0$ of the excited state e to a high vibrational level $v = v$ of the ground state g. The v level is (nearly) resonant with the $v' = 0$ level. The tunnelling rate depends on the vibrational wave function overlap of the two vibrational levels. Since the wave function of v has its highest amplitude at the turning points (see Fig. 2.4), the tunnelling rate is at a maximum if the minimum of the e parabola reaches the turning point, i.e. if ΔR is large

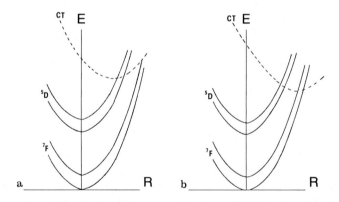

Fig. 4.9. The role of the charge-transfer state (CT) in the quenching of the luminescence of the Eu^{3+} ion. Only a few parabolas of the $4f^6$ configuration have been drawn. In (a), the situation for $Y_2O_3 : Eu^{3+}$ is depicted: the CT state feeds the emitting ^5D levels. In (b), the CT state has a larger offset. As a consequence the CT state populates, at least partly, the ground state levels, so that the luminescence is strongly reduced

Note that the useful properties of Y_2O_3–Eu^{3+} are based on a fast radiationless process, viz. the transition from the charge-transfer state to the excited levels of the $4f^6$ configuration. For crystalline GdB_3O_6–Eu^{3+} the same model holds. This composition

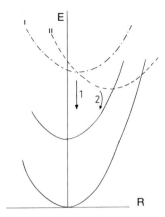

Fig. 4.10. Schematic configurational coordinate diagram for Pr^{3+} ($4f^2$). Drawn parabolas relate to the $4f^2$ configuration; the broken parabolas indicate two possible situations for the $4f^{5d}$ configuration (I and II). Excitation into I yields $d \rightarrow f$ emission from I (arrow 1). Excitation into II (with larger offset) yields a nonradiative transition to the $4f^2$ configuration (arrow 2) which may be followed by intraconfigurational $4f^2$ emission

can also be obtained as a glass. Interestingly enough, charge-transfer excitation in the glass results in Eu^{3+} luminescence with an efficiency which is an order of magnitude smaller than in the crystalline modification. This is also the case at 4.2 K. This observation has been ascribed to a larger offset of the charge-transfer parabola in the glass than in the crystal, so that in the glass the charge-transfer state empties mainly into the ^7F ground-state manifold. It is quite conceivable that the glass surroundings can counteract the expansion upon excitation less than the crystalline surroundings. The important consequence of this is that broad-band emission in glasses will have low efficiencies, unless the Stokes shift is small.

If the charge-transfer state moves to lower energy, the efficiency of the luminescence upon charge-transfer excitation decreases also. This is due to an increasing probability of the nonradiative transition from the charge-transfer state to the ^7F levels (compare the discussion on the difference between $CaWO_4$ and $CaMoO_4$ in this paragraph).

A different, but comparable, example is the Pr^{3+} ($4f2$) ion. There is an excited $4f5d$ configuration. Figure 4.10 gives the configurational coordinate diagram of Pr^{3+} in two different host lattices. If the offset of the $4f^{5d}$ state is small, radiative return to the $4f^2$ configuration has a higher probability than the nonradiative transition to the $4f^2$ configuration. If the offset is large, $4f^{5d}$ excitation leads to emission from the $4f^2$ configuration after a nonradiative $4f^5d \rightarrow 4f^2$ transition. The former situation is encountered for YBO_3, YOCl and LaB_3O_6, the latter for the apatite $Gd_{9.33}(SiO_4)_6O_2$ and Gd_2O_2S.

The Pr^{3+} case has an advantage over that of Eu^{3+}, viz. the higher excited state can emit, so that the Stokes shift can be measured. This gives information on the relaxation and the parabola offset. It was found that the nonradiative $4f^{5d} \to {}_4f^2$ transition becomes important if the Stokes shift is larger than 3000 cm^{-1}.

In conclusion, this section has summarized evidence in order to obtain some idea what are the factors determining the rate of radiationless processes in an isolated luminescent center, without going into any physical or mathematical details.

4.3 Efficiency

In the preceding section the expression "efficiency of the luminescence" was used frequently without giving a definition. A luminescent material which emits brilliantly is, of course, efficient. In this paragraph the several definitions of the efficiency of a luminescent material are given.

In the case of photoluminescence we distinguish the quantum efficiency (q), the radiant efficiency (η) and the luminous efficiency (L). The quantum efficiency q is defined as the ratio of the number of emitted quanta to the number of absorbed quanta. In the absence of competing radiationless transitions its value is 1 (or 100%). In the fundamental literature this is the important efficiency. The other two have a more technical importance.

The radiant efficiency η is defined as the ratio of the emitted luminescent power and the power absorbed by the material from the exciting radiation. The luminous efficiency L is the ratio of the luminous flux emitted by the material and the absorbed power.

Sometimes the term light output is used. This is the quantum efficiency multiplied by the amount of absorbed radiation. A high q does not necessarily imply a high light output: a photoluminescent material has only a high light output if the quantum efficiency and the optical absorption coeffient for the exciting wavelength are both high.

For cathode-ray (CR) excitation q is irrelevant. The radiant efficiency is defined as the ratio of the emitted power to the power of the electron beam falling on the luminescent material. This means that η_{CR} refers to the total power incident on the material, whereas η_{UV} refers to the power absorbed. The luminous efficiency for CR excitation is defined in a similar way as for photoexcitation. The radiant efficiency for X-ray excitation is defined as for CR-ray excitation.

In Table 4.5 we have gathered a few data on the several efficiencies of luminescent materials known to be efficient. These data were measured at room temperature [12]. The factors which restrict the photoluminescence efficiency, viz. the radiationless processes, were discussed above. For high-energy excitation, like cathode-ray or X-ray excitation, the situation is more complicated. Data like those presented in Table 4.5 for η_{CR} and n_X suggest that the maximum values of the radiant efficiency for host lattice excitation are restricted to values in between 10 and 20%. This in fact is the case. This observation has already drawn attention decades ago, and early explanations were offered in the sixties. The most detailed one is that by Robbins [13]. The next paragraph presents a summary of this work.

Table 4.5. Efficiencies of some luminescent materials at room temperature after Ref. [12]. See also text (Sect. 4.3)

Luminescent material	η^a (%)	q^a (%)	L^a (lm/W)	$\eta_{CR}{}^b$ (%)	$\eta_X{}^c$ (%)
MgWO$_4$	44	84	115		
Ca$_5$(PO$_4$)$_3$(F,Cl):Sb,Mn	34	71	125		
Zn$_2$SiO$_4$:Mn 6	35	70	175		
CaWO$_4$				3	6.5
ZnS:Ag				21	~ 20
Gd$_2$O$_2$S:Tb				11	13

[a] 250-270 nm excitation
[b] cathode-ray excitation
[c] X-ray excitation

4.4 Maximum Efficiency for High Energy Excitation [13]

In this section we will consider those types of exciting radiation which create many electrons and holes in the luminescent material under consideration. This type of excitation is usually indicated as high energy ionizing radiation. The most well-known examples are cathode rays and X rays, but one can also think of γ rays and α particles. It is usually assumed that the excitation process proceeds as follows:

The fast particle is absorbed in the lattice and creates secondary electrons and holes by ionization. One initial particle with high energy may create many electron-hole pairs. In CaWO$_4$, for example, one X-ray quantum may yield 500 quanta of emission. However, the fast particle and the created charge carriers may loose energy to the lattice by exciting vibrations. In order to create the average electron-hole pair (of energy E$_g$) a much larger amount of energy is needed. Long ago Shockley estimated this amount to be 3E$_g$ for semiconductors, i.e. the maximum possible conversion efficiency is on this ground reduced to $\frac{1}{3}$.

More generally we can write for the average energy required to create an electron-hole pair

$$E = \beta E_g. \tag{4.4}$$

Robbins [13] showed that β can even be larger than 3. The dependence of β on the so-called energy loss parameter K is given in Figure 4.11. The K parameter is given by

$$K \sim (\varepsilon_\sim^{-1} - \varepsilon_s^{-1})(h\nu_{LO})^{3/2}(1.5\ E_g)^{-1}. \tag{4.5}$$

Here $\varepsilon_\sim$ is the high-frequency dielectric constant, ε_s the static dielectric constant, and ν_{LO} the frequency of the longitudinal optical vibration mode. The values of β range from about 3 (GaP, ZnS, CsI, NaI), via 4 (La$_2$O$_2$S), 5.6 (Y$_3$Al$_5$O$_{12}$), to 7 (CaWO$_4$, YVO$_4$).

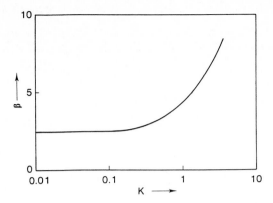

Fig. 4.11. The dependence of $\beta (= E/E_g)$ on the energy loss parameter K, defined in Eq. (4.5). See also text

The total expression for the radiant effiency looks as follows

$$\eta = (1 - r).\frac{h\nu_e}{E}.S.q. \tag{4.6}$$

Here r is the amount of radiation which is not absorbed, ν_e is the (averaged) frequency of the emitted radiation, E is defined by Eq. (4.4), S is the efficiency of transfer of electron-hole pair energy to the luminescent centre, and q is the quantum efficiency of the luminescent centre,

Assuming that all excitation energy is absorbed (i.e. $r = 0$), that all electron-hole pair energy arrives at the luminescent center (i.e. $S = 1$), and that $q = 100\%$, we arrive at the maximum radiant efficiency η_{max}:

$$\eta_{max} = h\nu_e E^{-1} = \frac{h\nu_e}{\beta E_g}. \tag{4.7}$$

The factor $h\nu_e$ accounts for the fact that the emitted energy will be less than the band gap energy. For example, in the case of ZnS : Ag $h\nu_e = 2.75$ eV and $E_g = 3.8$ eV, in the case of NaI : Tl 3.02 and 5.9, and of La_2O_2S : Eu 2.0 and 4.4, respectively. In this way the following η_{max} values can be calculated: $CaWO_4$ 8%, ZnS : Ag 25%, and Gd_2O_2S : Tb^{3+} 15%. The experimental values (see Table 4.5) are close to these values.

This discussion shows that high values of η_{max} may be expected if the optical vibrational modes are at low frequency (i.e. low β), and the emission is close to the band gap energy.

4.5 Photoionization and Electron-Transfer Quenching

In this section we will draw attention to another type of radiationless transition, viz. that as a consequence of photoionization. In some cases the luminescence of a center

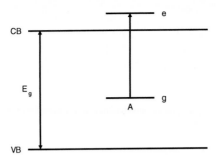

Fig. 4.12. The excited state e of the centre A lies in the conduction band of the host lattice. Photo-excitation of A (arrow) may be followed by photoionization

is strongly changed or even completely quenched by photoionization. The principle is illustrated in Figure 4.12. The luminescent center A has its ground state in the forbidden zone between valence and conduction bands. Its excited state lies in the conduction band. This implies that in the excited state an electron can easily be ionized from the center to the conduction band. It may recombine nonradiatively with a hole somewhere else, so that the luminescence is quenched. On the other hand the electron in the conduction band and the hole on the ionized centre will attract each other and may form an exciton. Since this exciton is bound to the luminescent center, radiative recombination of this exciton is known as impurity-bound exciton recombination [14]. However, the recombination may also be nonradiative.

A nice example is the isostructural series $CaF_2 : Yb^{2+}$, $SrF_2 : Yb^{2+}$ and $BaF_2 : Yb^{2+}$ [14]. The calcium compound shows normal Yb^{2+} emission (Sect. 3.3.3), the strontium compound shows a strikingly different emission, viz. impurity-bound exciton emission, and the barium compound does not show any emission at all. Obviously the Yb^{2+} energy level scheme moves upwards in the energy band picture of the host lattice going from CaF_2 to BaF_2.

Also $BaF_2 : Eu^{2+}$ does not show Eu^{2+} emission, but impurity-bound exciton emission. The same holds for $NaF : Cu^+$ where the excited state consists of a Jahn-Teller distorted Cu^{2+} ion which binds an electron. The absence of luminescence in $La_2O_3 : Ce^{3+}$ has also been ascribed to quenching by photoionization [15].

Closely related to photoionization and its consequences is quenching of luminescence by electron transfer. This process is well known in coordination chemistry, but has been overlooked in solid state studies. The principle is outlined in Figure 4.13 for a system of two species, A and B. The first excited state is one in which only A is excited ($A*+B$). At higher energies we find a charge-transfer state $A^+ + B^-$ with a large offset. Although $A^+ + B^-$ lies at higher energy than $A*+B$, judging from the absorption spectrum, the luminescence $A* \rightarrow A$ is quenched via the charge-transfer state.

This implies that a combination of a center which has a tendency to become oxidized with a center which has a tendency to become reduced will probably not show efficient luminescence. This is illustrated by some examples:

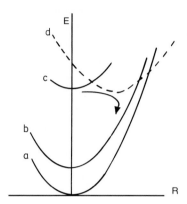

Fig. 4.13. Luminescence quenching by electron transfer. The ground state a consists of two species, A + B. In the excited states b and c the A ion is excited: A $*$ +B. State d is the electron-transfer state $A^+ + B^-$. Luminescence from the level c is quenched by electron transfer as indicated by the arrow

– whereas many rare earth ions show efficient luminescence in YVO_4, the ions Ce^{3+}, Pr^{3+} and Tb^{3+} do not. This is due to quenching via a charge-transfer state $(RE^{4+}+V^{4+})$ which is at low energy for these three ions and at much higher energy for the others.
– whereas many rare earth ions show efficient luminescence in cerium compounds, Eu^{3+} does not. Quenching occurs via a charge-transfer state $(Ce^{4+} + Eu^{2+})$ which is at low energy. Further examples are given in Ref. [16].

Note that quenching by electron transfer is a special case of quenching via a charge-transfer state as discussed above (Sect. 4.2.2) for Eu^{3+}. There is also no principal difference between quenching by electron transfer and quenching via photoionization. However, the former has a localized character, the latter is used in energy band models where delocalization is more important.

4.6 Nonradiative Transitions in Semiconductors

Finally nonradiative transitions which are specific for semiconductors are mentioned. In order to do so, we consider a specific radiative transition, viz. the donor-acceptor-pair emission (Sect. 3.3.9). In the excited state the donor and acceptor are occupied (Fig. 4.14). This excited luminescent center may show radiative and nonradiative transitions within the center itself (similar to the discussion in Sect. 4.2). In addition, however, there are other processes which are related to the valence and/or conduction band.

If the donor is thermally ionized (Fig. 4.14), the luminescence will be quenched unless it is trapped again at a donor site. A different nonradiative mechanism is

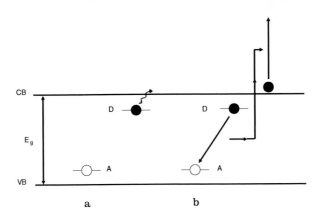

Fig. 4.14. Nonradiative transitions in a semiconductor. The donor-acceptor pair emission (DA) can be quenched by thermal ionization of one of the centers (*a*) or by an Auger process (*b*). In the latter case a conduction electron is promoted high into the conduction band

an Auger transition [17]. This is also illustrated in Figure 4.14. The energy of the excited donor-acceptor pair is used to excite a conduction band electron to a higher state in the conduction band. The hot conduction-band electron which has been created relaxes subsequently by intraband transitions. As a consequence the donor-acceptor pair emission is quenched.

More generally, an Auger transition can be defined as a transition in which energy is transferred from one electronic particle to another in such a way that in the final state the energy of one of the particles lies in a continuum. Auger processes can be classified as intrinsic or extrinsic. The former occur in the pure semiconductor, the latter involve electronic states of impurities like in the example in Figure 4.14. All types of luminescence transitions described in Section 3.3.9 can be quenched by Auger processes.

References

1. DiBartolo B (ed) (1980) Radiationless processes. Plenum, New York
2. DiBartolo B (ed) (1991) Advances in nonradiative processes in solids. Plenum, New York
3. Struck CW, Fonger WH (1991) Understanding luminescence spectra and efficiency using W_p and related functions, Springer, Berlin Heidelberg New York
4. Yen WM, Selzer PM (eds) (1981) Laser spectroscopy of solids. Springer, Berlin Heidelberg New York
5. Riseberg LA, Weber MJ ((1976) Progress in optics. In: Wolf E (ed) vol XIV. North-Holland, Amsterdam
6. van Dijk JMF, Schuurmans MFH (1983), J Chem Phys 78:5317
7. Berdowski PAM, Blasse G (1984) Chem Phys Letters 107:351
8. de Hair JThW, Blasse G (1976) J Luminescence 14:307; J Solid State Chem 19:263
9. Blasse G, van Vliet JPM, Verweij JWM, Hoogendam R, Wiegel M (1989) J Phys Chem Solids 50:583
10. Sabbatini N, Blasse G (1988) J Luminescence 40/41:288

11. Bleijenberg KC, Blasse G (1979) J Solid State Chem 28:303
12. Bril A (1962) in Kallman and Spruch (eds), Luminescence of organic and inorganic materials. Wiley, New York p 479; de Poorter JA, Bril A (1975) J Electrochem Soc 122:1086
13. Robbins DJ (1980) J Electrochem Soc 127:2694
14. Moine B, Courtois B, Pedrini C (1989) J Phys France 50:2105
15. Blasse G, Schipper W, Hamelink JJ (1991) Inorg Chim Acta 189:77
16. Blasse G, p 314 in ref. 1
17. Williams F, Berry DE, Bernard JE, p 409 in ref. 1

CHAPTER 5

Energy Transfer

5.1 Introduction

In Chapter 2, the luminescent center was brought into the excited state, whereas in Chapters 3 and 4 the return to the ground state was considered, radiatively and nonradiatively, respectively. In this chapter another possibility to return to the ground state is considered, viz. by transfer of the excitation energy from the excited centre (S^*) to another centre (A):

$S^* + A \rightarrow S + A^*$ (Figs 1.3 and 1.4).

The energy transfer may be followed by emission from A. Species S is then said to sensitize species A. However, A^* may also decay nonradiatively; in this case species A is said to be a quencher of the S emission.

Energy transfer between two centres requires a certain interaction between these centres. Nowadays the process of energy transfer is a well-understood phenomenon of which the more important aspects will be discussed here. For more profound treatments the reader is referred to the literature cited [1-3].

The organization of this chapter is as follows. In Sect. 5.2, we consider energy transfer between a pair of unlike luminescent centers. The theories of Förster and Dexter will be introduced. In Sect. 5.3, the topic is extended to energy transfer between identical centers. As a consequence of this, the phenomenon of concentration quenching of luminescence takes place. The section is split into two parts, one dealing with centers to which the weak-coupling scheme applies, the other dealing with centers to which the strong-coupling scheme applies. In Sect. 5.4, energy transfer in semiconductors is very briefly mentioned.

5.2 Energy Transfer Between Unlike Luminescent Centers

Consider two centers, S and A, separated in a solid by distance R (Fig. 5.1). We use the (classic) notation S and A (for sensitiser and activator); other authors use D and A (for donor and acceptor). The energy level schemes are also given in Fig. 5.1. An asterisk indicates the excited state. Let us assume that the distance R is so short that the centres S and A have a non-vanishing interaction with each other. If S is in the excited state and A in the ground state, the relaxed excited state of S may transfer

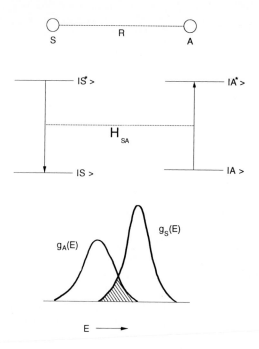

Fig. 5.1. Energy transfer between the centers S and A and an illustration of Eq. (5.1). The two centers are at a distance R (top). The energy level schemes and the interaction H_{SA} are given in the middle. The spectral overlap is illustrated at the bottom (hatched part)

its energy to A. The rate of such energy transfer processes has been calculated by Förster. Later Dexter extended this treatment to other interaction types.

Energy transfer can only occur if the energy differences between the ground- and excited states of S and A are equal (resonance condition) and if a suitable interaction between both systems exists. The interaction may be either an exchange interaction (if we have wave function overlap) or an electric or magnetic multipolar interaction. In practice the resonance condition can be tested by considering the spectral overlap of the S emission and the A absorption spectra. The Dexter result looks as follows:

$$P_{SA} = \frac{2\pi}{\hbar} |<S, A^*|H_{SA}|S^*, A>|^2 \cdot \int g_S(E) \cdot g_A(E) dE \qquad (5.1)$$

In Eq. (5.1) the integral presents the spectral overlap, $g_X(E)$ being the normalized optical line shape function of centre X (see Fig. 5.1, where the spectral overlap has been hatched). Equation [5.1] shows that the transfer rate P_{SA} vanishes for vanishing spectral overlap. The matrix element in Eq. (5.1) represents the interaction (H_{SA} being the interaction Hamiltonian) between the initial state $|S^*, A>$ and the final state $|S, A^*>$.

The distance dependence of the transfer rate depends on the type of interaction. For electric multipolar interaction the distance dependence is given by R^{-n} ($n = 6, 8, \ldots$

for electric-dipole electric-dipole interaction, electric-dipole electric-quadrupole inter-action, ...respectively). For exchange interaction the distance dependence is exponen-tial, since exchange interaction requires wave function overlap.

A high transfer rate, i.e. a high value of P_{SA}, requires a considerable amount of:
(1) resonance, i.e. the S emission band should overlap spectrally the A absorption band(s),
(2) interaction, which may be of the multipole-multipole type or of the exchange type. Only for some specific cases, is the interaction type known. The intensity of the optical transitions determines the strength of the electric multipolar interaction. High transfer rates can only be expected if the optical transitions involved are allowed electric-dipole transitions. If the absorption strength vanishes, the transfer rate for electric multipolar interaction vanishes too. However, the overall transfer rate does not necessarily vanish, because there may be contributions by exchange interaction. The transfer rate due to exchange interaction depends on wave function overlap (and, of course, spectral overlap), but not on the spectral properties of the transitions involved.

Over what distances can energy be transferred in this way? To answer this question it is important to realize that S* has several ways to decay to the ground state: energy transfer with a rate P_{SA}, and radiative decay with a rate P_S (the radiative rate). We neglect nonradiative decay (but it can be included in P_S). The critical distance for energy transfer (R_c) is defined as the distance for which P_{SA} equals P_S, i.e. if S and A are separated by a distance R_c, the transfer rate equals the radiative rate. For $R > R_c$ radiative emission from S prevails, for $R < R_c$ energy transfer from S to A dominates.

If the optical transitions of S and A are allowed electric-dipole transitions with a considerable spectral overlap, R_c may be some 30 Å. If these transitions are forbidden, we need exchange interaction for the transfer to occur. This restricts the value of R_c to some 5–8 Å.

If the spectral overlap consists of a considerable amount of overlap of an emission band and an allowed absorption band, there can be a considerable amount of radiative energy transfer: S* decays radiatively and the emitted radiation is reabsorbed. This can be observed from the fact that the emission band vanishes at the wavelengths where A absorbs strongly.

Energy transfer as described by Eq. (5.1) is nonradiative energy transfer. The occurrence of nonradiative energy transfer can be detected in several ways. If the excitation spectrum of the A emission is measured, the absorption bands of S will be found as well, since excitation of S yields emission from A via energy transfer. If S is excited selectively, the presence of A emission in the emission spectrum points to S → A energy transfer. Finally, the decay time of the S emission should be shortened by the presence of nonradiative energy transfer, since the transfer process shortens the life time of the excited state S*.

To obtain some feeling for transfer rates and critical distances, we will perform some simple calculations. We assume that the interaction is of the electric-dipolar type. In that case Eq. (5.1) together with $P_{SA}(R_c) = P_S$ yields for R_c the following expression [4]:

$$R_c^6 = 3 \times 10^{12} . f_A . E^{-4} . SO. \tag{5.2}$$

Table 5.1. Schematic energy transfer calculations between a broad-band and a narrow-line centre (see text)

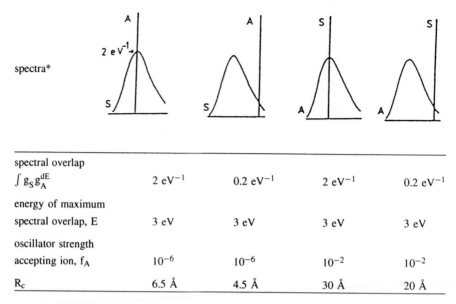

spectral overlap				
$\int g_S g_A^{dE}$	2 eV^{-1}	0.2 eV^{-1}	2 eV^{-1}	0.2 eV^{-1}
energy of maximum spectral overlap, E	3 eV	3 eV	3 eV	3 eV
oscillator strength accepting ion, f_A	10^{-6}	10^{-6}	10^{-2}	10^{-2}
R_c	6.5 Å	4.5 Å	30 Å	20 Å

* S: sensitizer (energy-transferring ion).
A: activator (energy-accepting ion). Height at band maximum 2 eV^{-1}. Band width 0.5 eV.

Here f_A presents the oscillator strength [1,5] of the optical absorption transition on A, E the energy of maximum spectral overlap, and SO the spectral overlap integral as given in Eq. (5.1). Equation (5.2) makes it possible to calculate R_c from spectral data.

In Table 5.1 the results of the calculation are presented. An S and an A center are considered, and R is always 4 Å. Transfer from a broad-band emitter to a narrow-line absorber is considered (the first two cases) as well as the reverse. The spectral overlap is optimal (1st and 3rd case) or marginal. For the narrow-line absorber, f_A is taken to be 10^{-6}, for the broad-band absorber 10^{-2} (forbidden and allowed transitions, respectively). E is taken to be 3 eV. These values are rather realistic, although lower f values may occur.

From Table 5.1 we can learn the following:

– energy transfer from a broad-band emitter to a line absorber is only possible for nearest neighbours in the crystal lattice
– transfer from a line emitter to a band absorber proceeds over fairly long distances.

Let us finally illustrate Eq. (5.1) by some examples:
(a) The Gd^{3+} ion shows energy transfer from the $^6P_{7/2}$ level to most rare earth ions, but not to Pr^{3+} and Tm^{3+}. Figure 2.14 shows that these two ions do not have energy

levels at the same energy as the $^6P_{7/2}$ level, so that the resonance condition is not satisfied, the spectral overlap vanishes, and the transfer rate becomes zero.

(b) In $Ca_5(PO_4)_3F:Sb^{3+},Mn^{2+}$ the Sb^{3+} ion can transfer energy to the Mn^{2+} ion. The Sb^{3+} emission covers several Mn^{2+} absorption transitions. These have very low f values (spin- and parity forbidden), and the transfer occurs by exchange interaction with $R_c \sim 7$ Å.

(c) In $Rb_2ZnBr_4:Eu^{2+}$ two types of Eu^{2+} ions are found with different spectra due to the presence of two types of crystallographic sites for Rb^+ in Rb_2ZnBr_4. Energy transfer occurs from the higher-emitting Eu^{2+} (415 nm) to the lower-emitting Eu^{2+} (435 nm). Since all optical transitions involved are allowed, it is not surprising that R_c has a high value (35 Å) [6].

5.3 Energy Transfer Between Identical Luminescent Centers

If we consider now transfer between two identical ions, for example between S and S, the same considerations can be used. If transfer between two S ions occurs with a high rate, what will happen in a lattice of S ions, for example in a compound of S? There is no reason why the transfer should be restricted to one step, so that we expect that the first transfer step is followed by many others. This can bring the excitation energy far from the site where the absorption took place: energy migration. If in this way, the excitation energy reaches a site where it is lost nonradiatively (a killer or quenching site), the luminescence efficiency of that composition will be low. This phenomenon is called concentration quenching. This type of quenching will not occur at low concentrations, because then the average distance between the S ions is so large that the migration is hampered and the killers are not reached.

Energy migration in concentrated systems has been an issue of research over the last two decades. Especially since lasers became readily available, the progress has been impressive. Here we will first consider the case that S is an ion to which the weak-coupling scheme applies. In practice this case consists of the trivalent rare earth ions. Subsequently we will deal with the case where S is an ion to which the intermediate- or strong-coupling scheme applies.

5.3.1 Weak-Coupling Scheme Ions

At first sight, energy transfer between identical rare earth ions seems to be a process with a low rate, because their interaction will be weak in view of the well-shielded character of the $4f$ electrons. However, although the radiative rates are small, the spectral overlap can be large. This originates from the fact that $\Delta R \simeq 0$, so that the absorption and emission lines will coincide. Further the transfer rate will easily surpass the radiative rate, since the latter is low. In fact energy migration has been observed in many rare earth compounds, and concentration quenching usually becomes effective for concentrations of a few atomic percent of dopant ions. Energy transfer over distances of up to some 10 Å is possible. As an example, we mention the transfer rate between Eu^{3+} ions or between Gd^{3+} ions which may be of the order of 10^7 s^{-1} if the distance is 4 Å or shorter. This has to be compared with a radiative rate of 10^2–10^3 s^{-1}. Consequently the excitation energy may be transferred more than 10^4 times during the life time of the excited state.

This type of research uses pulsed and tunable lasers as an excitation source. The rare earth ion is excited selectively with a laser pulse,and its decay is analyzed. The shape of the decay curve is characteristic of the physical processes in the compound under study. For a detailed review the reader is referred to the literature [1–3]. Here we give some results for specific situations. We assume that the object of our study consists of a compound of a rare earth ion S which contains also some ions A which are able to trap the migrating excitation energy of S by SA transfer.

(1) If excitation into S is followed by emission from the same S ion (i.e. the isolated ion case), or if excitation into S is followed after some migration by emission from S only, the decay is described by

$$I = I_o \exp(-\gamma t), \tag{5.3}$$

where I_o is the emission intensity at time $t = 0$, i.e. immediately after the pulse, and γ is the radiative rate. The decay is exponential. Equation (5.3) is identical to Eq. (3.3).

(2) If some SA transfer occurs, but no SS transfer at all, the S decay is given by

$$I = I_o \exp(-\gamma t - Ct^{3/n}), \tag{5.4}$$

where C is a parameter containing the A concentration (C_A) and the SA interaction strength, and $n \geq 6$ depending on the nature of the multipolar interaction. This decay is not exponential. Immediately after the pulse the decay is faster than in the absence of A. This is due to the presence of SA transfer. After a long time the decay becomes exponential with the radiative rate as a slope. This represents the decay of S ions which do not have A ions in the surroundings.

(3) If we allow for SS transfer, the situation becomes difficult. We consider first the extreme case that the rate of SS transfer (P_{SS}) is much higher than P_{SA} (fast diffusion). The decay rate is exponential and fast:

$$I = I_o \exp(-\gamma t) \exp(-C_A.P_{SA}.t). \tag{5.5}$$

(4) If $P_{SS} \ll P_{SA}$, we are dealing with diffusion-limited energy migration. For $t \to \infty$ the decay curve can be described by

$$I = I_o \exp(-\gamma t) \exp(-11.404 C_A.C^{1/4}.D^{3/4}.t) \tag{5.6}$$

if the sublattice of S ions is three dimensional. C is a parameter describing the SA interaction and D the diffusion constant of the migrating excitation energy. For lower dimensions non-exponential decays are to be expected.

In Fig. 5.2, we have given some of these decay curves. The temperature dependence of P_{SS} is very complicated. Due to inhomogeneous broadening, the S ions

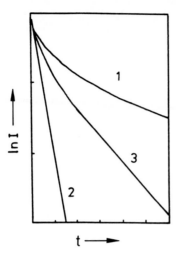

Fig. 5.2. Several possibilities for the decay curve of the excited S ion. The S emission intensity is plotted logarithmically versus time. Curve 1: no SS transfer (Eq. (5.3)); curve 2: rapid SS migration (Eq. (5.5)); curve 3: intermediate case (for example, Eq. (5.6))

are not exactly resonant, but their energy levels show very small mismatches. We need phonons to overcome these. At room temperature, the lines are broadened and phonons are available, so that these mismatches do not hamper the energy migration. However, at very low temperatures they hamper the energy migration, or make it even impossible. The theory of phonon-assisted energy transfer has been treated elsewhere [2]. For identical ions, one-phonon assisted processes are not of much importance. Two-phonon assisted processes have a much higher probability. One of these (a two-site nonresonant process) yields a T^3 dependence. Another (a one-site resonant process), which uses a higher energy level gives an $\exp(\frac{-\Delta E}{kT})$ dependence, where ΔE is the energy difference between the level concerned and the higher level. Figure 5.3 gives a schematical presentation of these two processes.

Let us now consider some examples. First we consider Eu^{3+} compounds. In $EuAl_3B_4O_{12}$, there is a three-dimensional Eu^{3+} sublattice with shortest Eu–Eu distance equal to 5.9 Å. At 4.2 K there is no energy migration at all, but at 300 K diffusion-limited energy migration occurs. The temperature dependence of P_{SS} is exponential with $\Delta E \sim 240$ cm^{-1}. This is due to the fact that the $^5D_0-^7F_0$ transition is forbidden under the relevant site symmetry (D_3), so that the multipolar interactions vanish. The distance of 5.9 Å is prohibitive for transfer by exchange interaction. At higher temperatures the $^7F_1$ level is thermally populated and multipolar interaction becomes effective. The experimental value of ΔE corresponds to the energy difference $^7F_0-^7F_1$ (compare Fig. 2.14). Analysis shows that the excited state makes 1400 jumps during its life time at room temperature with a diffusion length of 230 Å. Table 5.2 gives values of the transfer rate and the diffusion constant, together with those for other Eu^{3+} compounds.

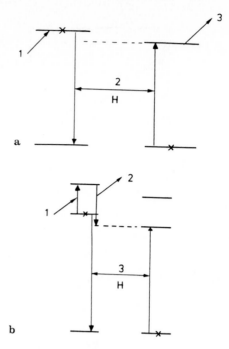

Fig. 5.3. A. Two-site nonresonant process in phonon-assisted energy transfer. (*1*) and (*3*) present the ion-phonon interaction, (*2*) the site-site coupling H. B. One-site resonant process in phonon-assisted energy transfer. (*1*) and (*2*) present the ion-phonon interaction, (*3*) the site-site coupling

Table 5.2. Energy migration characteristics in some Eu^{3+} compounds at 300 K (data from Ref. [7])

Compound	Shortest Eu–Eu distance	Diffusion constant $(cm^2 s^{-1})$	Hopping time [d] (s)
$EuAl_3B_4O_{12}$	5.9 Å	8×10^{-10}	8×10^{-7}
$NaEuTiO_4$	3.7 Å	2×10^{-8} [a]	2×10^{-8}
$EuMgB_5O_{10}$	4.0 Å	$\sim 10^{-8}$	$\sim 10^{-7}$
$Eu_2Ti_2O_7$	3.7 Å	9×10^{-12} [b]	3×10^{-5} [b]
		3×10^{-9} [c]	8×10^{-8} [c]
$Li_6Eu(BO_3)_3$	3.9 Å	2×10^{-9}	$\sim 10^{-7}$
$EuOCl$	3.7 Å	5.8×10^{-10}	4×10^{-7}

[a] $D = 8 \times 10^{-11}$ $cm^2 s^{-1}$ at 1.2 K.

[b] values at 15 K.

[c] values at 43 K.

[d] average time for one Eu^{3+}-Eu^{3+} transfer step.

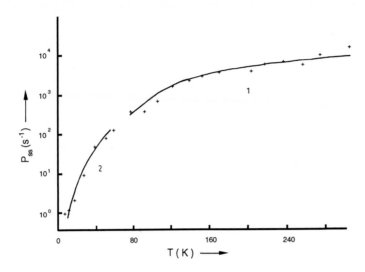

Fig. 5.4. Temperature dependence of the Eu^{3+}–Eu^{3+} transfer rate in $EuMgB_5O_{10}$. Line 1 is a fit using thermally activated migration via the $^7F_1$ level of the Eu^{3+} ion; line 2 is a fit to the T^3 temperature dependence predicted by a two-site nonresonant process

Samples of $EuAl_3B_4O_{12}$ which are so pure that the excited state does not reach a killer site during its life time, show efficient luminescence. Samples which contain a low concentration of killer sites do not show luminescence at 300 K. However, at 4.2 K they do, the migration being slowed down considerably or even completely. An example is a crystal of $EuAl_3B_4O_{12}$, grown from a K_2SO_4/MoO_3 flux [8]. These crystals contain ~ 25 ppm Mo^{3+} (on Al^{3+} sites). This ion is an efficient killer of the Eu^{3+} emission.

Two-dimensional energy migration was observed for $NaEuTiO_4$ and $EuMgAl_{11}O_{19}$, and one-dimensional energy migration for $EuMgB_5O_{10}$ and $Li_6Eu(BO_3)_3$. In the relevant crystal structures the Eu^{3+} ions form a two- and a one-dimensional sublattice, respectively.

The temperature dependence of P_{SS} is given for $EuMgB_5O_{10}$ in Fig. 5.4. At lower temperatures, we find the T^3 dependence expected for two-phonon assisted energy migration involving the $^7F_0$ and $^5D_0$ levels. At higher temperatures the temperature dependence becomes exponential, indicating transfer involving the $^7F_1$ and $^5D_0$ levels.

The situation in Eu^{3+} compounds can be characterized as follows:

– if the Eu–Eu distance is larger than 5 Å, exchange interaction becomes ineffective. Only multipolar interactions are of importance, and they will be weak anyhow. Actually, if sufficiently pure, $EuAl_3B_4O_{12}$ (Eu–Eu 5.9 Å), $Eu(IO_3)_3$ (Eu–Eu 5.9 Å), and $CsEuW_2O_8$ (Eu–Eu 5.2 Å) luminesce efficiently at 300 K.

– if the Eu–Eu distance is shorter than 5 Å, exchange interaction becomes effective. Examples are the intrachain migration in $EuMgB_5O_{10}$ and $Li_6Eu(BO_3)_3$ and the migration in Eu_2O_3 which is more rapid than in other compounds, even at very low temperatures.

For Tb^{3+} compounds, the situation is not essentially different, but the temperature dependence of the transfer rate shows another behavior, because the $^7F_6$ and $^5D_4$ levels are connected by an optical transition with a higher absorption strength than the $^7F_0$ and $^5D_0$ levels in the case of Eu^{3+}.

Recently, there has been a lot of interest in energy migration in Gd^{3+} compounds, because this opens interesting possibilities for obtaining new, efficient luminescent materials (see Chapter 6). The Gd^{3+} sublattice is sensitized and activated. The sensitizer absorbs efficiently ultraviolet radiation and transfers this to the Gd^{3+} sublattice. By energy migration in this sublattice the activator is fed, and emission results. Absorption and quantum efficiencies of over 90% have been attained. The physical processes can be schematically presented as follows:

$$\xrightarrow{\text{exc}} S \longrightarrow Gd^{3+} \xrightarrow{\text{nx}} Gd^{3+} \longrightarrow A \xrightarrow{\text{emission}}$$

Here nx indicates the occurrence of many Gd^{3+}–Gd^{3+} jumps. A suitable choice of S is Ce^{3+}, Bi^{3+}, Pr^{3+} or Pb^{2+}. For A there are many possibilities: Sm^{3+}, Eu^{3+}, Tb^{3+}, Dy^{3+}, Mn^{2+}, UO_6^{6-} and probably many more.

Not always is all of the excitation energy transferred. If only part of it is transferred, this is called cross-relaxation. Let us consider some examples. The higher-energy level emissions of Tb^{3+} and Eu^{3+} (Fig. 5.5) can also be quenched if the concentration is high. The following cross-relaxations may occur (compare Fig. 5.5):

$$Tb^{3+}(^5D_3) + Tb^{3+}(^7F_6) \rightarrow Tb^{3+}(^5D_4) + Tb^{3+}(^7F_0)$$

$$Eu^{3+}(^5D_1) + Eu^{3+}(^7F_0) \rightarrow Eu^{3+}(^5D_0) + Eu^{3+}(^7F_3).$$

The higher-energy level emission is quenched in favour of the lower energy level emission.

It is important to realize that we have met now two processes which will suppress higher-level emission, viz. multiphonon emission (Sect. 4.2.1) which is only of importance if the energy difference between the levels involved is less than about 5 times the highest vibrational frequency of the host lattice and which is independent of the concentration of the luminescent centres, and cross relaxation which will occur only above a certain concentration of luminescent centers since this process depends on the interaction between two centers.

Consider as an example the Eu^{3+} ion in YBO_3 and Y_2O_3. For low Eu^{3+} concentrations (say 0.1 mole %) we find in YBO_3 only $^5D_0$ emission, since the higher-level emissions are quenched by multiphonon emission (highest borate frequency ~ 1050 cm^{-1}). In Y_2O_3, however, such a low concentration of Eu^{3+} ions gives $^5D_3$, $^5D_2$, $^5D_1$ and $^5D_0$ emission (highest lattice frequency ~ 600 cm^{-1}). For 3% Eu^{3+} in Y_2O_3 the emission spectrum is dominated by the $^5D_0$ emission. The higher-level emission is quenched by cross relaxation in favour of the $^5D_0$ emission.

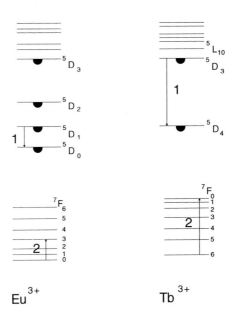

Fig. 5.5. Quenching of higher-level emission by cross relaxation. Left-hand side Eu^{3+}: the $^5D_1$ emission on ion 1 is quenched by transferring the energy difference $^5D_1$–$^5D_0$ to ion 2 which is promoted to the $^7F_3$ level. Right-hand side Tb^{3+}: the $^5D_3$ emission on ion 1 is quenched by transferring the energy difference $^5D_3$–$^5D_4$ to ion 2 which is promoted to the $^7F_0$ level

Up to now it has been demonstrated that concentration quenching of the luminescence of rare earth compounds consists of energy migration to killers in the case of Eu^{3+}, Gd^{3+}, Tb^{3+}. In the case of Sm^{3+} and Dy^{3+}, the above-mentioned cross relaxation is responsible for concentration quenching: the quenching of the luminescence occurs in ion pairs and not by migration. For other rare earth ions the situation is in between. We can use Pr^{3+} to illustrate the situation, and consider, in addition to energy migration, cross relaxation between Pr^{3+} ions, i.e. every Pr^{3+} ion can be a killer of its neighbours luminescence. Figure 5.6 gives possible cross relaxation processes which quench the $^3P_0$ and the $^1D_2$ emission of Pr^{3+}. The situation is even more complicated, because quenching may also occur in the isolated Pr^{3+} ion due to multi-phonon emission if the available frequencies are high enough. The energy gap below the $^3P_0$ level is 3500 cm^{-1}, that below the $^1D_2$ level 6500 cm^{-1} (see Fig. 5.6). Since the probability of all these processes depends strongly on the nature of the lattice and its constituents, the luminescence behaviour of Pr^{3+} compounds is expected to vary strongly from case to case. This is what has been observed.

Extensive research has been performed on the system $(La,Pr)F_3$. The $^3P_0$–$^1D_2$ nonradiative rate is very slow. At very low temperatures there is no energy migration due to small energy mismatches. Quenching of the $^3P_0$ emission occurs only by cross relaxation. At higher temperatures, however, energy migration among the $^3P_0$ level takes over. The Pr–Pr interaction is of the exchange type. The situation is different in

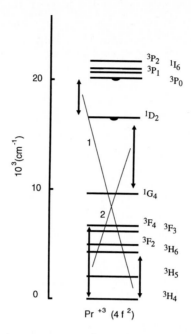

Fig. 5.6. The $^3P_0$ and the $^1D_2$ emission of Pr^{3+} can be quenched by cross relaxation as indicated by processes 1 and 2, respectively

a compound like PrP_5O_{14}. The high-energy vibrations of the phosphate group make the nonradiative $^3P_0 \rightarrow {}^1D_2$ transition the more probable process.

An interesting aspect of energy migration in concentrated rare-earth compounds is the influence of the magnetic order on the migration process. This has been studied on compounds of Gd^{3+} and Tb^{3+} by Jacquier et al. [9]. As an example we mention $GdAlO_3$ and $TbAlO_3$ which become antiferromagnetic at 3.9 K and 3.8 K, respectively. In the paramagnetic phase the experimental decays are in agreement with fast diffusion energy transfer. Below the Néel temperature, however, the decays are no longer exponential and considerably slower. The migration has become diffusion limited. The diffusion constants reported are 1.6×10^{-9} cm^{2}s^{-1} at 4.4 K for both compunds, but only 8×10^{-12} cm^{2}s^{-1} and 8×10^{-14} cm^{2}s^{-1} at 1.6 K for $GdAlO_3$ and $TbAlO_3$, respectively. In the antiferromagnetic phase the migration of excitation energy is slowed down, because nearest neighbour Gd^{3+} (or Tb^{3+}) ions are oriented antiparallel, which makes energy transfer by exchange interaction impossible. In $EuAlO_3$ such an effect does not occur. The Eu^{3+} ions (ground state $^7F_0$) do not carry a magnetic moment. This compound is another example of a Eu^{3+} compound in which energy migration occurs down to the lowest temperatures due to exchange interaction.

Let us now turn to energy migration in concentrated systems for which the weak-coupling scheme is no longer valid.

5.3.2 Intermediate- and Strong-Coupling Scheme Ions

Whether or not energy transfer will occur between identical ions with $S > 1$, depends in the first place on the spectral overlap of their emission and absorption spectra. The reader will feel intuitively that for ions with strongly Stokes-shifted emission $(S > 10)$ this spectral overlap will be very small or even vanish, so that energy transfer becomes impossible.

It is easy to make these feelings more quantitative [10]. Consider a system of two identical luminescent centers. The ground state of each center is denoted by $|g(i)v(i) >$, and the excited state by $|e(i)v'(i) >$. Here g and e indicate the electronic states (ground or excited state) and v and v' the vibrational states, and i numbers the centers, i.e. 1 or 2.

If H presents the interaction causing energy transfer, the transition matrix element is given by

$$M = < g(1)v(1), e(2)v'(2)|H|e(1)v'(1), g(2)v(2) > \tag{5.7}$$

If H operates on the electronic functions only,

$$M = < g(1)e(2)|H|e(1)g(2) > < v(1)|v'(1) > < v'(2)|v(2) > . \tag{5.8}$$

The transfer probability is proportional to M^2, i.e. to $|< v(1)|v'(1) >|^4$. At low temperatures the transfer transitions are restricted to the zero-vibrational levels, so that at 0K

$$M^2 \sim |< 0|0 >|^4. \tag{5.9}$$

Here $|0>$ denotes the zero-vibrational level. At low temperatures, therefore, the transfer probability vanishes if the spectra do not show vibrational structure with a zero-phonon line.

A clear example is $CaWO_4$. The excitation energy on the tungstate group remains localized at that center in spite of nearby tungstate groups. However, $< 0|0 >$ is practically zero, so that P_{SS} vanishes.

At higher temperatures the occupation of vibrational levels is no longer restricted to the zero-vibrational levels. The spectral bands broaden, and, as a consequence the spectral overlap may increase enough to make (thermally stimulated) energy transfer possible. We will now discuss several examples for illustration [7,10].

The first case to be considered here is the luminescence of complexes with U^{6+} as the central ion. Their spectra consist of zero-vibrational transitions followed by a rich vibrational structure (see Fig. 3.5). Examples are UO_2^{2+}, octahedral UO_6^{6-}, trigonal prismatic UO_6^{6-} and tetrahedral UO_4^{2-}.

The presence of a zero-phonon line implies that the spectral overlap does not vanish, so that transfer between two identical species may occur if the interaction is strong enough. This has actually been observed. Several uranates have been studied and it was shown that the absorbed excitation energy migrates over the uranium lattice. At low temperatures this energy is usually captured by optical traps, i.e. species with a slightly different crystal field, and therefore different energy level diagram.

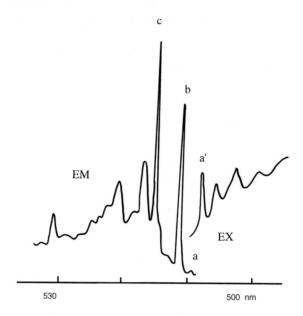

Fig. 5.7. Part of the emission and excitation spectrum of Ba_2CaUO_6 at 4.2 K. The zero-phonon line in the excitation spectrum is indicated by a'. There is no corresponding line in the emission spectrum. The lines a, b and c in the emission spectrum are the zero-phonon lines of the emission spectra belonging to three different uranate centres. This shows that excitation of the intrinsic uranate centres (with line a') is followed by energy migration to the extrinsic or defect uranate centres (with lines a, b and c)

The emission originates from these traps, so that the zero-phonon line in emission is at slightly lower energy than that in absorption (see Fig. 5.7). The energy difference between these two zero-phonon lines is equal to the depth of these traps. At higher temperatures the traps are no longer effective, because they are too shallow. In uranates the luminescence is then usually quenched due to energy migration to U^{5+} centers. The latter act as quenching centers. In uranyl compounds, which are closer to stoichiometry, often strong intrinsic uranyl emission is observed, the quenching center concentration being too low to capture a substantial part of the excitation energy. Examples are $Cs_2UO_2Cl_4$ and $UO_2(NO_3)_2.6H_2O$.

Another intermediate coupling case is the Bi^{3+} ion in $Cs_2Na(Y,Bi)Cl_6$. The spectra show zero-vibrational transitions accompanied by many vibronic transitions. Energy migration among the Bi^{3+} ions occurs even at low Bi^{3+} concentrations. For $Cs_2NaBiCl_6$ the intrinsic Bi^{3+} emission is no longer observed. Only a red broad-band emission appears. This has been ascribed to a Bi^{3+} ion next to impurity ions (like oxygen). This center is fed by energy migration over the Bi^{3+} ions followed by transfer to the impurity centers (trapping).

However, as we have seen above, a completely different situation may appear in other Bi^{3+} compounds (see also Sect. 3.3.7). In $Bi_4Ge_3O_{12}$, for example, excitation into a Bi^{3+} ion is followed by a pronounced relaxation, resulting in an emission with a large Stokes shift (strong-coupling scheme). As a consequence, the relaxed excited state is out of resonance with the neighbouring ions, so that the excitation energy remains on the excited Bi^{3+} ion and energy migration does not occur.

The Ce^{3+} ion is another case where the Huang-Rhys parameter may vary considerably, so that the Stokes shift varies, and energy migration may or may not occur. The optical transitions are of the $f-d$ type and as such allowed. In $CeBO_3$ the Ce^{3+} emission at 300 K is quenched by energy migration to killer sites. The spectral overlap between absorption and emission is large. In CeF_3, however, the spectral overlap is much smaller. As a consequence no $Ce^{3+}-Ce^{3+}$ energy transfer occurs: the emission of CeF_3 is not quenched at room temperature due to a stronger relaxation. In fact CeF_3 is proposed for important scintillator applications (see Chapter 9). The same situation occurs for $CeMgAl_{11}O_{19}$, an important host lattice for a green lamp phosphor, viz. $CeMgAl_{11}O_{19}-Tb$. If a Ce^{3+} compound does not show energy migration among the Ce^{3+} sublattice (like CeF_3 or $CeMgAl_{11}O_{19}$), large quantities of Tb^{3+} are necessary to quench the Ce^{3+} emission by $Ce^{3+}-Tb^{3+}$ transfer. This is quite understandable, since the range of the $Ce^{3+}-Tb^{3+}$ transfer is restricted.

Another case of strong coupling is formed by groups like tungstate and vanadate (see Sects. 2.3.2 and 3.3.5). These are oxidic anions with a central metal ion which has lost all its d electrons. Examples are WO_4^{2-}, WO_6^{6-}, VO_4^{3-}, MoO_4^{2-}. Usually the Stokes shift of their emission is so large ($\sim 16\,000$ cm^{-1}), that energy migration is completely hampered, even at room temperature. A well known example is $CaWO_4$. However, in other cases thermally-activated energy migration occurs, e.g. in YVO_4, Ba_2MgWO_6 and Ba_3NaTaO_6. In these compounds the Stokes shift is considerably smaller ($\sim 10\,000$ cm^{-1}). This makes YVO_4-Eu^{3+} a very efficient red phosphor: excitation into the vanadate group is followed by energy migration over the vanadate groups to Eu^{3+} centers. Pure YVO_4 emits only weakly at room temperature. However, if the temperature is lowered, or the V^{5+} concentration lowered by P^{5+} substitution, the migration is hampered, YVO_4 starts to emit efficiently, and YVO_4-Eu^{3+} yields blue vanadate emission upon vanadate excitation.

This "dilution experiment", in which V^{5+} is replaced by P^{5+}, has no consequences in the case of $CaWO_4$ and $YNbO_4$, for example. The composition $CaSO_4:W$ shows exactly the same luminescence as $CaWO_4$, and $YTaO_4:Nb$ the same as $YNbO_4$. This proves that the luminescent WO_4^{2-} and NbO_4^{3-} groups in $CaWO_4$ and $YNbO_4$, respectively, are to be considered as isolated luminescent centers in spite of the short distance to their nearest neighbours. This is due to the larger relaxation in the excited state in the cases of niobate and tungstate relative to vanadate, as is immediately clear from the larger Stokes shift ($16\,000$ vs $10\,000$ cm^{-1}). The fundamental reason for this fact is, however, not yet clear.

Up till here, it has been assumed in this chapter that free charge carriers do not play a role. This is different in the case of semiconductors.

5.4 Energy Transfer in Semiconductors

If excitation of a luminescent material results in the creation of free charge carriers, additional forms of energy transfer will take place. It is here not the place to consider these in detail (see, for example, Refs [11-13]).

Energy absorbed by the formation of free charge carriers can be transported through the crystal lattice. It should be realized that electrons and holes must move into the same direction (ambipolar diffusion). Therefore, the energy transport is determined by the carriers which have the shorter lifetime and the smaller diffusion constant. As discussed above (Sect. 3.3.9), the charge carriers will be trapped at centers in the crystal lattice where radiative recombination is one of the possibilities of returning to the ground state.

Energy transfer can also occur by excitons. An exciton is an excited state of the crystal lattice in which an electron and a hole are bound and can propagate through the lattice [14]. They can be divided into two classes, viz. Frenkel and Wannier excitons. In the former the electron-hole separation is of the order of an atomic radius, i.e. it can be considered as a localized excitation. In the latter this separation is large in comparison with the lattice constant. Consequently, its binding energy is much smaller than that of the Frenkel exciton.

Examples of Frenkel excitons were met above (YVO_4, $CaWO_4$, $Bi_4Ge_3O_{12}$). An extreme case is solid Kr with an exciton binding energy of ~ 2 eV and a radius of ~ 2 Å. Semiconductors are examples in which Wannier excitons occur (e.g. Ge, GaAs, CdS, TlBr). As an extreme example we mention InSb with an exciton binding energy of ~ 0.6 meV and a radius of ~ 600 Å. Such an exciton is only stable at very low temperatures [15].

Free excitons can be bound to defects or become self trapped. In both cases the electron and hole will recombine, either radiatively or nonradiatively. An example was discussed in Sect. 3.3.1. It will be clear that energy transfer by excitons is of a general importance and occurs in semiconductors as well as in insulators.

The contents of Chapters 2–5 form the fundamental background knowledge which is necessary for a discussion of luminescent materials (Chapters 6–10). The reader should realize, however, that the treatment has been kept as simple and restricted as possible.

References

1. Henderson B, Imbusch GF (1989) Optical spectroscopy of inorganic solids. Clarendon, Oxford
2. Yen WM, Selzer PM (eds) (1981) Laser spectroscopy of solids, Topics in Applied Physics 49. Springer, Berlin Heidelberg New York
3. Di Bartolo B (ed) (1984) Energy transfer processes in condensed matter. Plenum, New York
4. Blasse G (1987) Mat Chem Phys 16 (1987) 201; (1969) Philips Res Repts 24:131
5. Atkins PW (1990) Physical chemistry, 4th edn, Oxford University Press, Oxford
6. Blasse G (1986) J Solid State Chem 62:207

7. Blasse G (1988) Progress Solid State Chem 18:79
8. Kellendonk F, Blasse G (1981) J Chem Phys 75:561
9. Salem Y, Joubert MF, Linarrs C, Jacquier B (1988) J Luminescence 40/41:694
10. Powell RC, Blasse G (1980) Structure and Bonding 42:43
11. Bernard JE, Berry DE, Williams F, p 1 in Ref. [3]
12. Klingshirn C, p 285 in Ref. [3]
13. Broser I (1967) In: Aven M, Prener JS (eds) Physics and chemistry of II-VI compounds. North Holland, Amsterdam, Chapter 10
14. See e.g. C. Kittel, Introduction to solid state physics, Wiley, New York, several editions
15. Sturge MD (1982) In: Rashba ÉI, Sturge MD (eds) Excitons. North Holland, Amsterdam, Chapter 1

Lamp Phosphors

6.1 Introduction

The previous chapters presented an outline of the phenomenon of luminescence in solids. They form the background for the following chapters which discuss luminescent materials for several applications, viz. lighting (Chapter 6), television (Chapter 7), X-ray phosphors and scintillators (Chapters 8 and 9), and other less-general applications (Chapter 10). These chapters will be subdivided as follows:

- the principles of the application
- the preparation of the materials
- the luminescent materials which were or are in use or have a strong potential to become used; a discussion of their luminescence properties in terms of Chapters 2–5
- problems in the field.

The emphasis will be on the materials in view of the topic of this book.

6.2 Luminescent Lighting [1-3]

Luminescent lighting started even before the Second World War*. The ultraviolet radiation from a low-pressure mercury discharge is converted into white light by a phosphor layer on the inner side of the lamp tube. These lamps are much more efficient than the incandescent lamp: a 60 W incandescent lamp yields 15 lm/W, a standard 40 W luminescent lamp 80 lm/W.

A luminescent lamp is filled with a noble gas at a pressure of 400 Pa, containing 0.8 Pa mercury. In the discharge the mercury atoms are excited. When they return to the ground state, they emit (mainly) ultraviolet radiation. About 85% of the emitted radiation is at 254 nm and 12% at 185 nm. The remaining 3% is found in the longer wavelength ultraviolet and visible region (365, 405, 436 and 546 nm).

---

* We use the term luminescent lighting instead of the generally used fluorescent lighting, since most of the luminescent materials that are used do not show fluorescence (which is defined as an emission transition without spin reversal, i.e. $\Delta S = 0$; see also Appendix III).

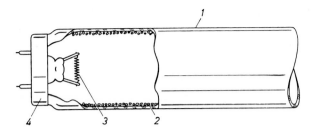

Fig. 6.1. Cross section of a low-pressure luminescent lamp. *1* glass tube; *2* luminescent powder; *3* cathode; *4* lamp cap

The lamp phosphor converts the 254 and 185 nm radiation into visible light (Fig. 6.1). It is in direct contact with the mercury discharge which rules out many potential candidates. For example, sulfides cannot be used in lamps since they react with mercury. A lamp phosphor should absorb the 254 and 185 nm radiation strongly and convert the absorbed radiation efficiently, i.e. their quantum efficiency should be high.

A luminescent lighting lamp has to emit white light, so that the sun, our natural lighting source, is imitated. The sun is a black body radiator, so that its emission spectrum obeys Planck's equation:

$$E(\lambda) = \frac{A\lambda^{-5}}{\exp(B/T_c) - 1} \qquad (6.1)$$

Here A and B are constants, λ the emission wavelength and T_c the temperature of the black body. With increasing T_c the color of the radiator moves from (infra)red into the visible. In luminescent lamp terminology, "white" is used for 3500 K light, "cool-white" for 4500 K, and "warm-white" for 3000 K.

According to the principles of colorimetry, each color can be matched by mixing three primary colors. It is possible to represent colors in a color triangle [2]. Most currently used is the chromaticity diagram standardized by the Commission Internationale d'Eclairage. It is depicted in Fig. 6.2. For a definition of the color coordinates x and y, see Refs. [2] and [3]. The real colors cover an area enclosed by the line representing the spectral colors and the line connecting the extreme violet and the extreme red. The points within this area represent unsaturated colors.

The color points corresponding to Eq. (6.1) are given by the black body locus (BBL). Colors lying on the BBL are considered to be white. White light can be generated in different ways. The simplest one is to mix blue and orange. However, it is also possible to mix blue, green and red. Blending a number of emission bands into a continuous spectrum also yields, of course, white light. All these examples of color mixing are used in lamps, as we will see below.

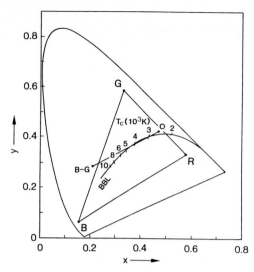

Fig. 6.2. CIE chromaticity diagram with black body locus (BBL). See also text. Reproduced with permission from Ref. [3]

Apart from the color point, there is another important lamp characteristic, viz. the color rendition. This property depends on the spectral energy distribution of the emitted light. It is characterized by comparing the color points of a set of test colors under illumination with the lamp to be tested and with a black body radiator. The color rendering index (CRI) equals 100 if the color points are the same under illumination with both sources. Under illumination with a lamp with low CRI, an object does not appear natural to the human eye.

In addition to the low-pressure mercury lamp discussed above, there is the high-pressure mercury lamp (Fig. 6.3). The gas discharge is contained in a small envelope surrounded by a larger bulb. The phosphor coating is applied to the inside of the outer bulb, so that there is no contact with the discharge.

In the high-pressure lamp the discharge also shows strong lines at 365 nm. The ideal phosphor for this lamp should, therefore, not only absorb short-wavelength ultraviolet radiation, but also long-wavelength. Further this discharge shows a considerable amount of blue and green emission. However, it is deficient in red. The phosphor has to compensate for this deficiency, so that it should have a red emission.

The phosphor temperature in the high-pressure lamp increases to 300°C, so that the emission should have a very high quenching temperature.

Since high-pressure lamps are used for outdoor lighting, the requirements for color rendering are less severe than for low-pressure lamps. However, if the phosphor is left out, red objects appear to be dull brown: this not only makes human skin look terrible, but also finding a red car in a parking lot problematic.

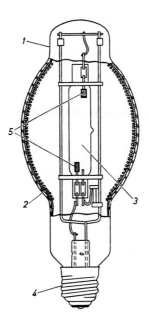

Fig. 6.3. Cross section of a high-pressure luminescent lamp. *1* glass bulb; *2* luminescent powder; *3* quartz envelope for gas discharge; *4* lamp cap; *5* electrodes

6.3 The Preparation of Lamp Phosphors

The lamp bulbs are coated with phosphor by using a suspension of phosphor powder particles. A lamp phosphor is therefore prepared as a powder. In principle this is done by standard solid state techniques in which intimate mixtures of starting materials are fired under a controlled atmosphere [4]. As a simple example we consider $MgWO_4$: it is prepared by mixing basic magnesium carbonate and tungsten trioxide in open silica crucibles at about 1000°C. Much more complicated is the case of the calcium halophosphate phosphor $Ca_5(PO_4)_3(F,Cl)$: Sb,Mn which is made by firing a mixture of $CaCO_3$, $CaHPO_4$, CaF_2, NH_4Cl, Sb_2O_3 and $MnCO_3$. Actually the history of the preparation of this material is a beautiful illustration that increasing control and knowledge yields results: the light output of this phosphor has increased considerably during a long period of time. For more details the reader is referred to Chapter 3 in Ref. [2].

The luminescent activator concentration is of the order of 1%. Therefore high-quality starting materials and a clean production process are prerequisite for obtaining luminescent materials with a high efficiency. The controlled atmosphere is necessary to master the valence of the activator (for example Eu^{2+} or Eu^{3+}) and the stoichiometry of the host lattice. Also the particle-size distribution of the phosphor needs to be controlled; this depends on the specific material under consideration.

In order to obtain homogeneous phosphors it is often necessary to leave the simple solid state technique. Coprecipitation may be of importance, especially if the activator and the host lattice ions are chemically similar. This is, of course, the case with rare-earth activated phosphors. For example, $Y_2O_3 : Eu^{3+}$ can be prepared profitably by coprecipitating the mixed oxalates from solution and firing the precipitate [5]. Actually the mixed oxides have become available commercially.

Usually phosphors decline slowly during lamp life [2]. This can be due to several processes:

- photochemical decomposition by 185 nm radiation from the mercury discharge (an illustrative approach to this problem is given in Ref. [6])
- reaction with excited mercury atoms from the discharge
- diffusion of sodium ions from the glass.

Quite often, coarse phosphors appear to be more stable than fine-grained phosphors. Obviously a high specific surface makes the phosphors more sensitive to interaction with radiation, mercury, and so on. This does not come as a surprise.

6.4 Photoluminescent Materials

6.4.1. Lamp Phosphors for Lighting
6.4.1.1. Early Phosphors

In the early period of luminescent lighting (1938–1948), a mixture of two phosphors was used, viz. $MgWO_4$ and $(Zn,Be)_2SiO_4 : Mn^{2+}$. The tungstate has a broad bluish-white emission band with a maximum near 480 nm (Fig. 6.4) and can be efficiently excited with short wavelength ultraviolet radiation. The emission spectrum of $(Zn,Be)_2SiO_4 : Mn^{2+}$ is given in Fig. 6.5. It covers the green to red part of the visible spectrum.

The phosphor $MgWO_4$ is an example of a luminescent material with 100% activator concentration, since each octahedral tungstate group in the lattice is able to luminesce. However, there is no concentration quenching. This is due to the large

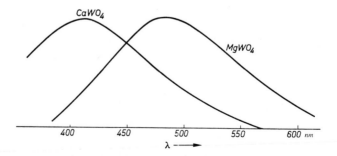

Fig. 6.4. Emission spectra of $MgWO_4$ and $CaWO_4$

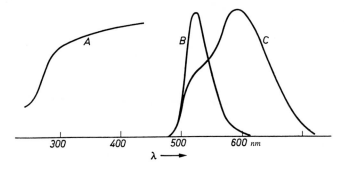

Fig. 6.5. Emission spectra of $Zn_2SiO_4 : Mn^{2+}$ (*B*) and $(Zn,Be)_2SiO_4 : Mn^{2+}$ (*C*). Curve *A* gives the diffuse reflection spectrum of $Zn_2SiO_4 : Mn^{2+}$

Stokes shift of the emission which brings the relaxed emitting state out of resonance with the neighbors. This, in turn, can be related to the nature of the optical transition which is a charge-transfer transition in the tungstate group. Therefore ΔR in Fig. 2.3 is large. This yields not only a strongly Stokes-shifted emission, but also a very broad emission which is of importance for the color rendering.

The broadness of the $(Zn,Be)_2SiO_4 : Mn^{2+}$ emission is due to another reason. Actually the emission of beryllium-free $Zn_2SiO_4 : Mn^{2+}$ is narrow (Fig. 6.5). Let us, therefore, start with the latter phosphor which shows a bright-green emission.

Both Zn_2SiO_4 and Be_2SiO_4 have the phenacite structure with all metal ions in four coordination. The Mn^{2+} $(3d^5)$ ion has therefore also four coordination. All optical transitions within the $3d^5$ configuration are spin- and parity forbidden (see Sect. 2.3.1). As a consequence excitation into these transitions does not yield a high light output. However, the luminescence of the Mn^{2+} ion in Zn_2SiO_4 shows a strong excitation band in the 250 nm region which is probably due to a charge-transfer transition. Anyway, excitation into this band yields a high light output of the Mn^{2+} emission. The emission transition is $^4T_1 \rightarrow {}^6A_1$. The emission is relatively narrow, because the transition occurs within a given electron configuration (viz. $3d^5$).

If we replace part of the Zn^{2+} ions by Be^{2+} ions, the crystal field on the Mn^{2+} ions will vary from ion to ion depending on the nature of the neighboring metal ions. This is due to the large difference between the ionic radii of Zn^{2+} and Be^{2+} (0.60 Å and 0.27 Å, respectively). Therefore, the emission band broadens relative to that of $Zn_2SiO_4 : Mn^{2+}$. Obviously the introduction of Be^{2+} increases the crystal field on the Mn^{2+} ions, so that the emission shifts to longer wavelength (Fig. 2.10).

A serious drawback of this Mn^{2+} phosphor is its poor maintenance in a lamp. It easily picks up mercury from the gas discharge and is liable to decompose under ultraviolet radiation. In addition beryllium is highly toxic [7], and nowadays not acceptable for application. In 1948, these phosphors were replaced by one phosphor with blue and orange emission, viz. Sb^{3+}- and Mn^{2+}-activated calcium halophosphate.

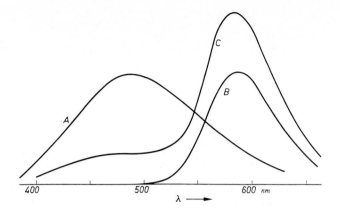

Fig. 6.6. Emission spectra of calcium halophosphates. *A*: Sb^{3+} emission; *B*: Mn^{2+} emission; *C*: warm-white halophosphate

6.4.1.2 The Halophosphates [1-3]

The halophosphates with composition $Ca_5(PO_4)_3X$ (X = F, Cl) are closely related to hydroxy-apatite, the chief constituent of bones and teeth. The apatite crystal structure is hexagonal and offers two different sites for calcium, viz. Ca_I and Ca_{II}. The Ca_I sites form linear columns. Each Ca_I site is trigonally prismatic coordinated by six oxygen ions with average Ca_I–O distance 2.42 Å. Individual prisms share top and bottom. The sides of the prisms are capped by oxygen at a Ca_I-O distance of 2.80 Å, so that the total coordination is nine (by oxygen) and zero (by halogen). The Ca_{II} site has one halogen neighbour (Ca_{II}–F: 2.39 Å) and six oxygens (average Ca_{II}–O: 2.43 Å).

The optical absorption edge of the pure host lattice is at about 150 nm: all excitation energy from the mercury discharge has to be absorbed by the activators. Peculiarly enough, the crystallographic position of the Sb^{3+} and Mn^{2+} ions in this lattice is not exactly known. Optical and electron-paramagnetic-resonance data have shown that Mn^{2+} has a preference for the Ca_I site [8]. It is generally assumed that Sb^{3+} is on Ca_{II} sites with oxygen on the neighboring halogen site for charge compensation [8]: $(Sb^{\bullet}_{Ca}.O'_F)^x$ in the Kröger notation [9]. However, this has been doubted by Mishra et al. [10], who suggest antimony on a phosphorus site with an oxygen vacancy for charge compensation: $(Sb''_P.V^{\bullet\bullet}_O)^x$. This in turn has been questioned [11]. This situation shows that the apatite structure is a complicated one.

Jenkings et al. [12] discovered that Sb^{3+}-doped calcium halophosphate is a very efficient blue-emitting phosphor under 254 excitation (see Fig. 6.6). The Sb^{3+} ion has $5s^2$ configuration and its $^1S_0 \rightarrow {}^3P_1$ and $^1P_1$ absorption bands are situated at 255 and 205 nm, respectively (Fig. 6.7) [11]. The Stokes shift of the emission is enormous, viz. 19 000 cm^{-1} at room temperature.

These data teach us to become a little suspicious about the phantastic properties of the halophosphate phosphor. Excitation is into the $^1S_0 \rightarrow {}^3P_1$ transition. This is a spin-forbidden transition, and although it has a certain absorption strength by spin-orbit

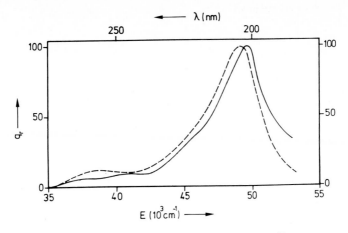

Fig. 6.7. Excitation spectrum of the Sb^{3+} emission of $Ca_5(PO_4)_3F:Sb^{3+}$ at 4.2 K (continuous curve) and 290 K (broken curve)

coupling, it will never be as strongly absorbing as an allowed transition. Figure 6.7 shows this clearly, because the output for $^1S_0 \rightarrow {}^1P_1$ excitation is nearly an order of magnitude larger than for $^1S_0 \rightarrow {}^3P_1$ excitation (see also Sect. 2.3.5). In addition, the large Stokes shift predicts a low quantum efficiency (Sect. 4.2.2.). Actually $q \simeq 70\%$ (Sect. 4.1). Such large Stokes shifts in case of s^2 ions were discussed in Sect. 3.3.7.

When the halophosphate host lattice contains not only Sb^{3+} but also Mn^{2+}, part of the energy absorbed by the Sb^{3+} ions is transferred to Mn^{2+}. The Mn^{2+} ion shows an orange emission (see Fig. 6.6). The 254 nm radiation of the mercury discharge is hardly absorbed by the Mn^{2+} ions. The critical distance for energy transfer from Sb^{3+} to Mn^{2+} is about 10 Å; the acting mechanism is exchange [8]. This is not unexpected, since all optical transitions of the Mn^{2+} ion in the visible region are strongly forbidden (Sect. 5.2).

By carefully adjusting the ratio of the Sb^{3+} and Mn^{2+} ion concentrations, a white-emitting phosphor can be obtained with color temperatures ranging between 6500 and 2700 K. Figure 6.6 presents the emission spectrum of a warm-white halophosphate. A large drawback of the halophosphate lamps is the fact that it is impossible to have simultaneously high brightness and high color rendering: if the brightness is high (efficacy ~ 80 lm/W), the color rendering index (CRI) is of the order of 60; the CRI value can be improved up to 90, but then the brightness decreases (~ 50 lm/W) [13]. The use of rare-earth activated phosphors has made it possible to achieve the combination of a high efficacy (~ 100 lm/W) with a high CRI value (~ 85).

6.4.1.3 Phosphors for the Tricolor Lamp

It was predicted by Koedam and Opstelten [14] that a luminescent lamp with an efficacy of 100 lm/W and a CRI of 80-85 can be obtained by combining three phophors which emit in narrow wavelength intervals centered around 450, 550 and 610 nm. A few years later such a lamp was realized using rare-earth activated phosphors [13]. This type of lamp is known as the tricolor lamp.

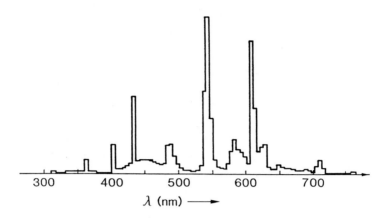

Fig. 6.8. Emisson spectrum of a tricolor lamp with a color temperature of 4000 K. Reproduced with permission from Ref. [3]

As we have seen before, rare-earth ions can have narrow band or line emission (Chapters 2 and 3), and as a matter of fact Eu^{3+} is the best candidate for the red-emitting component and Tb^{3+} for the green-emitting component. They show line emission. For the blue the Eu^{2+} ion is taken with a narrow blue emission band, considerably narrower than the Sb^{3+} emission band.

The chromaticity diagram of Fig. 6.2 shows how white light can be obtained from the halophosphate phosphor by blending blue-green (BG) with orange (O), and how it can be obtained using three phosphors, i.e. by blending blue (B), green (G), and red (R). Figure 6.8 shows the emission spectrum of a tricolor lamp with a color temperature of 4000 K.

The individual phosphors for the tricolor lamp will now be discussed in separate sections.

6.4.1.4 Red-Emitting Phosphors

The material $Y_2O_3 : Eu^{3+}$ fulfills all the requirements for a good red-emitting phosphor. Its emission is located at 613 nm and all other emission lines are weak. It can be easily excited by 254 nm radiation, and its quantum efficiency is high, viz. close to 100%.

The excitation of $Y_2O_3 : Eu^{3+}$ has already been discussed in Sect. 2.1: the 254 nm radiation is absorbed by the charge-transfer transition of the Eu^{3+} ion, the 185 nm radiation by the host lattice. Obviously the charge-transfer state is situated in the configurational coordinate diagram in such a way that it feeds the emitting levels exclusively (see Fig. 6.9). The emission spectrum of Eu^{3+} was discussed in Sect. 3.3.2.

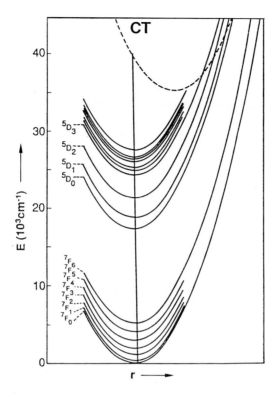

Fig. 6.9. Configuration coordinate diagram for Eu^{3+} in Y_2O_3. *CT*: charge-transfer state. Reproduced with permission from Ref. [38]

It consists of the $^5D_0 - ^7F_J$ line emissions with that to $J = 2$ dominating due to its hypersensitivity (see Fig. 6.10).

However, the actual situation is more complicated. In the first place Y_2O_3 offers two sites to the Eu^{3+} ion, one with C_2 and one with S_6 symmetry (see Fig. 6.11). There are three times more C_2 sites, and Eu^{3+} is assumed to occupy these two types of sites in a statistical way. The S_6 site has inversion symmetry, so that the Eu^{3+} ion on this site will only show the $^5D_0 - ^7F_1$ magnetic-dipole emission (Sect. 3.3.2) which is situated around 595 nm. The strongly forbidden character of the $^5D_0 - ^7F_J$ transitions of $Eu^{3+}(S_6)$ becomes clear from the value of the decay time of 8 ms compared with 1.1 ms for $Eu^{3+}(C_2)$ [15].

The unwanted $^5D_0 - ^7F_1$ emission of the S_6 site is suppressed in the commercial 3% Eu^{3+} samples by the occurrence of energy transfer from $Eu^{3+}(S_6)$ to $Eu^{3+}(C_2)$. The critical distance for this transfer amounts to about 8 Å and exchange as well as dipole-quadrupole interaction seem to play a role [15].

Other advantages of the 3% concentration are the quenching of the unwanted higher level emission from Eu^{3+} (i.e. $^5D_J - ^7F$, $J > 0$; see Sect. 5.3) by cross-relaxation,

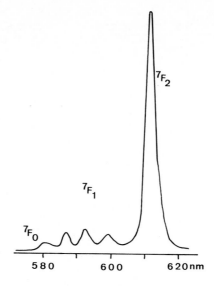

Fig. 6.10. Emission spectrum of $Y_2O_3 : Eu^{3+}$. The final levels of the $^5D_0 - ^7F_J$ transitions are indicated

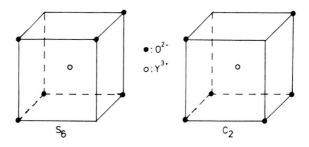

Fig. 6.11. Oxygen surroundings of $Y^{3+}(Eu^{3+})$ in $Y_2O_3 : Eu^{3+}$. On the left hand side is the S_6 site with inversion symmetry, on the right hand side the C_2 site

and a sufficient absorption strength of the 254 nm radiation. On the other hand the high europium concentration makes the phosphor very expensive, especially since high purity Y_2O_3 is required which is also not cheap. Impurities in Y_2O_3 tend to act as competing absorbing centers, i.e. they absorb the 254 nm radiation without converting it into visible light. A serious impurity is iron. It has been estimated that 5 ppm Fe lowers the quantum efficiency by 7% [16]. Below we will discuss research performed in order to lower the price of the red-emitting phosphor.

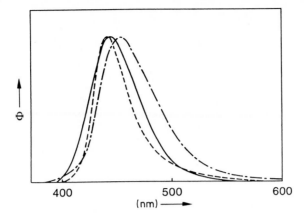

Fig. 6.12. Emission spectra of three blue-emitting Eu^{2+} phosphors. Full line: $BaMgAl_{10}O_{17}$: Eu^{2+}; broken line: $Sr_5(PO_4)_3Cl:Eu^{2+}$; broken line with dots: $Sr_2Al_6O_{11}:Eu^{2+}$. Reproduced with permission from Ref. [3]

6.4.1.5 Blue-Emitting Phosphors

The highest lamp output is expected for a blue-emitting phosphor with an emission maximum at 450 nm, whereas the best CRI is found with an emission maximum at 480 nm. Since the tricolor lamp aims at high light output in combination with good color rendition, only phosphors with an emission maximum between 440 and 460 nm are of practical interest. Figure 6.12 shows the emission spectrum of three Eu^{2+}-activated phosphors which satisfy the requirements: $BaMgAl_{10}O_{17}:Eu^{2+}$, $Sr_3(PO_4)_5Cl:Eu^{2+}$, and $Sr_2Al_6O_{11}:Eu^{2+}$ [3]. Their quantum efficiencies are about 90%. The spectroscopy of the Eu^{2+} ion was treated in Sect. 3.3.3b.

The compound $BaMgAl_{10}O_{17}$ has a crystal structure related to the magnetoplumbite structure. This structure consists of spinel layers with interlayers containing Ba^{2+}. The complicated nature of this composition is illustrated by the fact that originally it was written as "$BaMg_2Al_{16}O_{27}$" [13]. The situation has been discussed by Smets et al. [17].

The compound $Sr_5(PO_4)_3Cl$ belongs to the halophosphates mentioned above, whereas $Sr_2Al_6O_{11}$ is built up of alternating layers of AlO_4 tetrahedra and of AlO_6 octahedra [18]. The structure is shown schematically in Fig. 6.13.

6.4.1.6 Green-Emitting Phosphors

The green-emitting ion in the tricolor lamp is Tb^{3+}. Its first allowed absorption band is $4f^8 \rightarrow 4f^75d$ (Sect. 2.3.4.). It often lies at too high energy to make 254 nm excitation effective. In order to absorb the 254 nm radiation efficiently, a sensitizer has to be used. For this purpose the Ce^{3+} ion is very suitable. Its $4f \rightarrow 5d$ transition is situated at lower energy than the corresponding $4f^8 \rightarrow 4f^75d$ transition of Tb^{3+}. Table 6.1

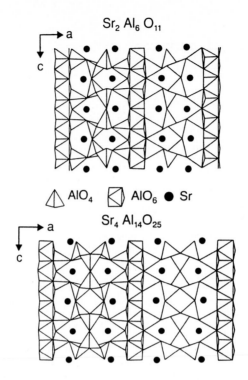

Fig. 6.13. Crystal structures of $Sr_2Al_6O_{11}$ and $Sr_4Al_{14}O_{25}$. Reproduced with permission from Ref. [3]

Table 6.1. Quantum efficiency q for ultraviolet (UV) and visible (VIS) emission for 254 nm excitation of green phosphors applied in the tricolor lamp [3].

Composition	q_{UV} (%)	q_{VIS} (%)
$Ce_{0.67}Tb_{0.33}MgAl_{11}O_{19}$	5	85
$Ce_{0.45}La_{0.40}Tb_{0.15}PO_4$	7	86
$Ce_{0.3}Gd_{0.5}Tb_{0.2}MgB_5O_{10}$	2	88

shows the green phosphors in use, with their chemical composition and quantum efficiencies in the ultraviolet and visible.

The host lattice $CeMgAl_{11}O_{17}$ has the magnetoplumbite structure, $LaPO_4$ the monazite structure, and $GdMgB_5O_{10}$ a structure consisting of a two-dimensional framework of BO_3 and BO_4 groups in which the Mg^{2+} ions are in octahedral coordination and Gd^{3+} in ten coordination [19]. The Gd^{3+} polyhedra form isolated zig-zag chains. The shortest Gd–Gd intrachain distance is about 4 Å, the shortest Gd–Gd interchain distance 6.4 Å.

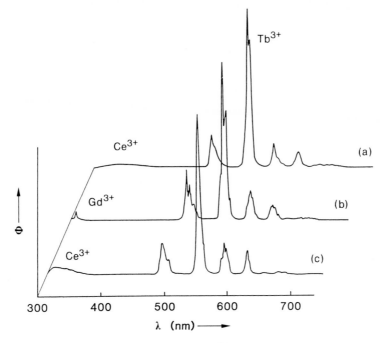

Fig. 6.14. Emission spectra of three green-emitting Tb^{3+} phosphors. (a): $CeMgAl_{11}O_{19}:Tb^{3+}$; (b): $(Ce,Gd)MgB_5O_{10}:Tb^{3+}$; (c): $(La,Ce)PO_4:Tb^{3+}$. Reproduced with permission from Ref. [3]

Figure 6.14 shows the emission spectra of the three green phosphors. In all of them some ultraviolet emission originating from Ce^{3+} or Gd^{3+} is present. The energy transfer phenomena are different as will be discussed now.

The Ce^{3+} emission of $CeMgAl_{11}O_{19}$ has its emission maximum at 330 nm and the first excitation maximum at 270 nm, so that the Stokes shift is large (~ 8000 cm^{-1}) [20]: energy transfer between Ce^{3+} ions does not occur and concentration quenching is absent. The $Ce^{3+} \rightarrow Tb^{3+}$ transfer is a one-step transfer process. Since this transfer is restricted to neighbors due to the forbidden nature of the intraconfigurational $4f^8$ transitions on Tb^{3+}, high Tb^{3+} concentrations are necessary to quench the Ce^{3+} emission. Even 33% Tb is not able to do so completely (see Table 6.1).

In $(La,Ce)PO_4$ the Stokes shift of the emission is smaller (~ 6000 cm^{-1}), and the emission of $CePO_4$ is partly concentration quenched. Energy migration over the Ce^{3+} ions assists in the transfer to Tb^{3+} and less Tb^{3+} is needed. Nevertheless the UV output is relatively high (see Table 6.1). This is due to the fact that the $Ce^{3+}–Ce^{3+}$ transfer is more efficient than the $Ce^{3+}–Tb^{3+}$ transfer. At the shortest internuclear rare-earth distance the transfer rates amount to 10^{11}s^{-1} and 3×10^8s^{-1}, respectively. This phosphor system has been discussed at length recently [21].

In $GdMgB_5O_{10}:Ce^{3+},Tb^{3+}$ the situation is even more complicated. Excitation with 254 nm radiation occurs in the Ce^{3+} ion which transfers its energy to the Gd^{3+} ions.

Table 6.2. Energytransfer processes in the green phosphors for the tricolor lamp.

Phosphor	Energy transfer processes
$CeMgAl_{11}O_{19}$: Tb	$Ce^{3+} \rightarrow Tb^{3+}$
$LaPO_4$: Ce,Tb	$Ce^{3+} \rightarrow Ce^{3+}$ *
	$Ce^{3+} \rightarrow Tb^{3+}$
$GdMgB_5O_{10}$: Ce, Tb	$Ce^{3+} \rightarrow Gd^{3+}$
	$Gd^{3+} \rightarrow Gd^{3+}$ *
	$Gd^{3+} \rightarrow Tb^{3+}$

* This transfer occurs many times repeatedly.

In order to have efficient transfer, the Ce^{3+} emission in the absence of Gd^{3+} must be situated at about 280 nm which makes transfer to the 6_1 levels of Gd^{3+} possible. This restricts the Stokes shift of the Ce^{3+} emission to some 4000 cm^{-1}. This has been observed in a few cases only, because not only the Stokes shift should be small, but also the first absorption transition of the Ce^{3+} ion should be at relatively high energy, requiring an ionic host lattice (see Sect. 2.2). Host lattices which satisfy these requirements are GdB_3O_6, GdF_3, $NaGdF_4$ (but not $LiGdF_4$) and $GdMgB_5O_{10}$ [22]. If the Ce^{3+} levels shift to somewhat lower energy (for example in $Li_6Gd(BO_3)_3$: Ce^{3+}), the transfer occurs only in the opposite direction, i.e. from Gd^{3+} to Ce^{3+}. Other ions can also be used as sensitizer of the Gd^{3+} sublattice viz. Pr^{3+}, Pb^{2+}, Bi^{3+} [22].

In $GdMgB_5O_{10}$: Ce,Tb practically all Ce^{3+} excitation energy is transferred to the Gd^{3+} ions. Subsequently, the energy migrates over the Gd^{3+} sublattice (see Sect. 5.3). It is trapped by the emitting Tb^{3+} ions. Care should be taken that the energy is not trapped by impurity ions which act as killers. Their role can be overruled by using a high Tb^{3+} concentration.

Finally we note that the energy migration over the Gd^{3+} sublattice seems to be less one-dimensional than the crystal structure suggests [23]. Table 6.1 shows that the use of Gd^{3+} host lattices makes it possible to convert more ultraviolet radiation into visible light. Table 6.2 summarizes the energy transfer processes in the green phosphors considered above.

6.4.1.7 Phosphors for Special Deluxe Lamps

Tricolor lamps show only emission in restricted wavelength intervals. For objects with a reflection spectrum peaking outside these regions the color appearance under illumination with a tricolor lamp will differ from the one under illumination with a black body radiator. Although a CRI of 85 quarantees a normal appearance for most objects, some typical colors will look unnatural under illumination with a tricolor lamp. For certain applications, therefore, a higher CRI is required. Examples of such applications are museum illumination and flower displays. For this purpose special deluxe lamps were developed with a CRI of 95. Simultaneously we have to accept an efficacy drop to 65 lm/W [3].

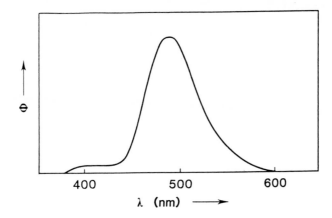

Fig. 6.15. Emission spectrum of $Sr_4Al_{14}O_{25}:Eu^{2+}$

As pointed out above (Sect. 6.4.1.5), a higher CRI can be obtained by using a blue-emitting phosphor with an emission maximum at 490 nm. A further increase of the CRI can be obtained by using band instead of line phosphors for the red and green. In this way a more or less continuous spectrum extending from the blue to the red is obtained.

A suitable blue emission is given by $Sr_4Al_{14}O_{25}:Eu^{2+}$ with an emission maximum at 490 nm (see Fig. 6.15) and a quantum efficiency of 90%. The crystal structure of $Sr_4Al_{14}O_{25}$ is given in Fig. 6.13 and is related to that of $Sr_2Al_6O_{11}$ [3]. In addition to the 490 nm emission band there is a weak emission band at 410 nm. This is due to the presence of two different crystallographic sites for Sr in $Sr_4Al_{14}O_{25}$, so that there are also two types of Eu^{2+} ions. Although both sites have the same abundancy, the 490 nm emission is by far the dominating emission. This favourable situation is due to efficient energy transfer from the 410 nm emitting Eu^{2+} ion to the 490 nm emitting Eu^{2+} ion. There is a large spectral overlap between the 410 nm emission band and the excitation band of the 490 nm emitting Eu^{2+} ion, and all optical transitions involved ($4f^7 \leftrightarrow 4f^65d$) are allowed. Therefore all conditions for efficient energy transfer are fullfilled (compare Sect. 5.2). A critical distance of some 35 Å has been found [24].

A broad-band red-emitting phosphor has been obtained in the host lattice $GdMgB_5O_{10}$. Again Ce^{3+} is used as a sensitizer, but this time Mn^{2+} as an activator. The Mn^{2+} ion on Mg^{2+} sites, i.e. in octahedral coordination, yields a red emission peaking around 630 nm. This emission corresponds to the $^4T_1 \rightarrow {}^6A_1$ transition (Sect. 3.3.4.c). The transfer to Mn^{2+} occurs via the Gd^{3+} sublattice as described above (Sect. 6.4.1.6). By using a composition $Ce_{0.2}Gd_{0.6}Tb_{0.2}Mg_{0.9}Mn_{0.1}B_5O_{10}$ a phosphor is obtained which emits simultaneously green and red.

An efficient broad-band green-emitting phosphor is not known, but can be simulated by combining Tb^{3+} emission with the Mn^{2+} emission of the halophosphate phosphor. A lamp containing a mixture of $Sr_2Al_6O_{11}:Eu^{2+}$, $GdMgB_5O_{10}:Ce^{3+},Tb^{3+},Mn^{2+}$ and $Ca_5(PO_4)_3(F,Cl):Sb^{3+},Mn^{2+}$ has a CRI of 95 and an efficacy of 65 lm/W. Its emission spectrum is shown in Fig. 6.16.

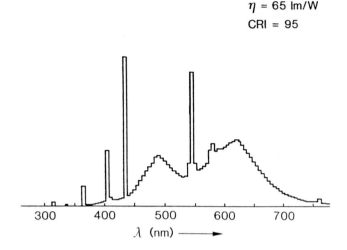

Fig. 6.16. Emission spectrum of a Special Deluxe lamp with a color temperature of 4000 K. Reproduced with permission from Ref. [3]

Figure 6.16 shows on the short wavelength side the blue mercury lines. These can be efficiently suppressed by adding $Y_3Al_5O_{12} : Ce^{3+}$. This phosphor, with garnet structure, absorbs blue light and converts it with high efficiency into yellow emission (see Fig. 6.17). The optical transitions involved are due to Ce^{3+}. In the garnet structure this ion undergoes such a strong crystal field, that the lowest $4f \rightarrow 5d$ transition is in the visible (Sect. 2.3.4). A comparison of Ce^{3+} in $Y_3Al_5O_{12}$ with Ce^{3+} in $GdMgB_5O_{10}$ (Sect. 6.4.1.6) illustrates how large the influence of the host lattice on the energy levels of a luminescent ion may be.

6.4.1.8 The Maintenance of Phosphors in Tricolor Lamps

The phosphors in the tricolor lamp show another advantage over the halophosphate phosphors, viz. a much better maintenance during lamp life. In Fig. 6.18 the output decrease after 2000 hours of burning is plotted as a function of the wall load. The higher stability of the rare-earth activated phosphors is translated into a higher maintenance. The value of the wall load is determined by the tube diameter (see Fig. 6.18). Tricolor and Special Deluxe lamps are now available in a tube diameter of 25 mm, to be compared with the 36 mm of the halophosphate tube. It even proved to be possible to reduce the tube diameter to 10 mm. With this small diameter the discharge tube can be fold up and the compact luminescent lamp is born.

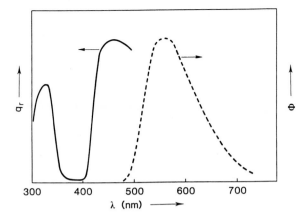

Fig. 6.17. Emission (dashed line) and excitation (continuous line) spectra of the luminescence of $Y_3Al_5O_{12}:Ce^{3+}$

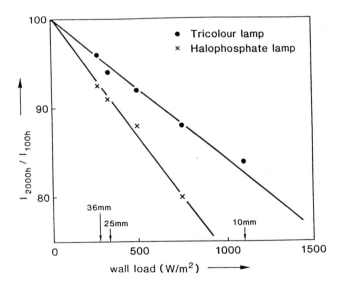

Fig. 6.18. Ratio of the lamp outputs after 2000 and 100 hours of burning as a function of the wall load. Crosses are for a halophosphate lamp, circles for a tricolor lamp. Commercial lamp diameters are indicated. Reproduced with permission from Ref. [3]

6.4.2 Phosphors for Other Lamp Applications

Low-pressure mercury discharge lamps are not only used for lighting. Since it is in principle possible to have a phosphor with every wavelength desired, there are many, more specialized applications. Here we mention a few.

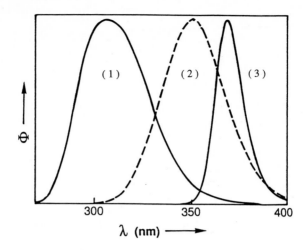

Fig. 6.19. Emission spectra of several ultraviolet-emitting phosphors. (*1*): $SrAl_{12}O_{19}:Ce^{3+}$, Mg^{2+}; (*2*): $BaSi_2O_5:Pb^{2+}$; (*3*): $SrB_4O_7:Eu^{2+}$

In phototherapy the emission of the lamp should correspond to the stimulation spectrum of the human skin. For $\lambda <\sim 300$ nm sunburn (erythema) occurs, and for $\lambda > 330$ nm direct pigmentation (skin darkening) occurs, but this disappears after a short time [3]. In the intermediate range delayed pigmentation (melanin production) occurs yielding a long lasting tan. Figure 6.19 shows the emission spectra of three ultraviolet-emitting phosphors.

It is clear that $SrAl_{12}O_{19}:Ce^{3+},Mg^{2+}$ (with magnetoplumbite structure and Ce^{3+} on Sr^{2+} sites and Mg^{2+} on Al^{3+} sites for charge compensation) will induce severe sunburn. On the other hand $SrB_4O_7:Eu^{2+}$, discussed in Sect. 3.3.3b, will induce only direct pigmentation. For a long lasting tan it is necessary to use lamps with $BaSi_2O_5:Pb^{2+}$. Considerations of this type are of importance in the development of sun-tanning lamps.

The Gd^{3+} emission is used in lamps to control psoriasis [3]. This skin disease cannot be cured, but by using ultraviolet treatment it can be controlled. The 312 nm emission of Gd^{3+} controls effectively the spread of psoriasis without too much sunburn. However, the Gd^{3+} ion cannot be excited directly with 254 nm radiation, since all optical transitions within its $4f^7$ configuration are strongly forbidden and the excited $4f^65d$ configuration is at very high energy ($> 70\,000$ cm^{-1}). Therefore it is necessary to use a sensitizer. For these lamps $GdBO_3:Pr^{3+}$ or $(La,Gd)B_3O_6:Bi^{3+}$ can be used [3].

In $GdBO_3:Pr^{3+}$ the 254 nm excitation is absorbed by Pr^{3+} ($4f^2 \rightarrow 4f5d$). After relaxation the Pr^{3+} ion transfers its excitation energy to Gd^{3+}. Energy migration over the Gd^{3+} sublattice follows. Since the Pr^{3+} ion has no energy levels coinciding with the lowest excited level of the Gd^{3+} ion, the excitation energy is emitted by Gd^{3+} itself. This is shown schematically in Fig. 6.20.

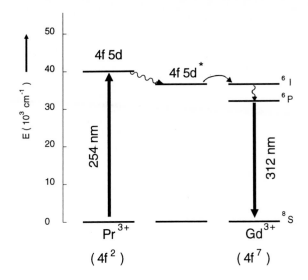

Fig. 6.20. The luminescence of $GdBO_3 : Pr^{3+}$. Excitation is into the $4f^2 \rightarrow 4f5d$ transition of Pr^{3+}, which ion transfers to the Gd^{3+} ion, which emits its ^6P emission. The level $4f5d^*$ indicates the relaxed excited state of Pr^{3+}. Energy migration over the Gd^{3+} sublattice is omitted for simplicity

In $(La,Gd)B_3O_6 : Bi^{3+}$ the transfer phenomena are similar. The Bi^{3+} ion ($6s^2$ configuration) absorbs the 254 nm radiation in the $^1S_0 \rightarrow {}^3P_1$ transition (Sect. 2.3.5). The relaxed $^3P_1$ state transfers this excitation energy to the Gd^{3+} ion from which emission occurs.

Further, blue-emitting $Sr_2P_2O_7 : Eu^{2+}$, with an emission maximum at 420 nm and $q \sim 90\%$, can be used in lamps in the phototherapy of hyperbilirubinemia (an excess of bilirubin in the blood serum which may result in permanent brain damage in newborn infants).

6.4.3 Phosphors for High-Pressure Mercury Vapour Lamps

Phosphors for this application should have a red emission which can be excited by long- as well as short-wavelength ultraviolet radiation with high quantum efficiency up to 300°C (see Sect. 6.2). Here we mention three different phosphors which can be used, viz. magnesium(fluoro) germanate doped with Mn^{4+}, $(Sr,Mg)_3(PO_4)_2 : Sn^{2+}$, and (modified) $YVO_4 : Eu^{3+}$.

The formula of the first host lattice has been given as $Mg_{28}Ge_{7.5}O_{38}F_{10}$. The Mn^{4+} ion ($3d^3$ configuration) absorbs over the whole ultraviolet range with an intense charge-transfer transition. The emission is in the deep red (620–670 nm), consists of several lines, and is due to the $^2E \rightarrow {}^4A_2$ transition (Sect. 3.3.4b). Thermal quenching occurs only above 300°C as is to be expected for a narrow line emission. The decay

Table 6.3. Emission lines of the Mn^{4+} activated fluorogermanate [2] and their assignment (see also Fig. 6.22).

Spectral position		Assignment (values in cm^{-1})
nm	cm^{-1}	
626.2[a]	15.969[a]	$0-0+405$ (ν_4) b
632.5[a]	15.810[a]	$0-0+246$ (ν_3) c
642.5	15.564	$0-0$
652.5	15.326	$0-0-238$ (ν_3) c
660.0	15.152	$0-0-412$ (ν_4) b
670.0	14.925	$0-0-639$ (ν_1?)

a lacking at 77 K

b intensity ratio of anti-Stokes and Stokes lines at 300 K amounts to 0.1 (experimental) and 0.13 (calculated).

c intensity ratio 0.4 (exp.) and 0.30 (calc.)

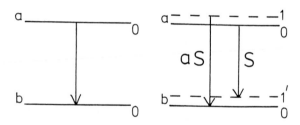

Fig. 6.21. Schematic representation of a zero-phonon transition (left-hand side), and the Stokes (*S*) and anti-Stokes (*aS*) vibronic transitions (right-hand side) in emission. Lettering refers to electronic states, numbering to vibrational states

time is long, viz. $\sim$ 3.5 ms, as is to be expected for a spin- and parity-forbidden transition.

Since the spectroscopy of this phosphor is incorrectly described in the book on lamp phosphors [2], we add here, also as an illustration of the theory, a few comments on the spectroscopy. In view of its electron configuration (d^3), the Mn^{4+} ion will be octahedrally coordinated. The emission lines are tabulated in Table 6.3. There is a zero-phonon transition (Sect. 2.1) which at low temperatures is followed by vibronic lines due to coupling with the asymmetric Mn^{4+}–O^{2-} deformation and stretching modes, ν_4 and ν_3, respectively. These uneven modes relax the parity selection rule. At room temperature there occur also anti-Stokes vibronics (Figs. 6.21 and 6.22). The vibrational modes in the excited state and ground state are equal within the experimental accuracy as is to be expected for the narrow $^2E \rightarrow {}^4A_2$ transition [25,26]. The intensity ratio of the Stokes and anti-Stokes vibronic lines agrees with the Bose-Einstein distribution [26].

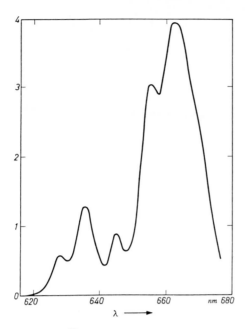

Fig. 6.22. Emission spectrum of Mn^{4+}-activated magnesium fluorogermanate

As argued in Sect. 4.2, the position of the charge-transfer state determines the quenching temperature. In the present case this means that the ^2E level will decay nonradiatively via the charge-transfer state. This has been suggested before in a discussion of the Mn^{4+} emission in $CaZrO_3$ [27]. In this composition the charge-transfer state is at $30\,000$ cm^{-1}, yielding a quenching temperature of 300 K. In the fluorogermanate these values are $35\,000$ cm^{-1} and 700 K, respectively. This illustrates how important the position of the charge-transfer state is for efficient luminescence.

Butler [2] has shown that, when carefully prepared, Sn^{2+}-activated phosphors can be efficient phosphors. For example, $(Sr,Mg)_3(PO_4)_2 : Sn^{2+}$ shows a broad-band red emission with a maximum at about 630 nm and a thermal quenching which starts above 300°C. The Sn^{2+} ion has $5s^2$ configuration, but its luminescence is not so simple as the discussion in Sect. 3.3.7 may suggest. For example, it is not easy to understand why the emission is so far in the visible, whereas in many other host lattices it is in the ultraviolet or blue. Donker et al. have found that the coordination number is of critical influence: high coordination yields ultraviolet or blue emission, low coordination yellow or red emission [28].

Of a more recent date are compositions based on $YVO_4 : Eu^{3+}$ which was originally introduced as the red primary in color television tubes (see Chapter 7) [29]. This is a very interesting phosphor material. It absorbs the greater part of the ultraviolet spectrum in the VO_4^{3-} group. By energy migration over the vanadate groups the Eu^{3+} ion is reached (Sect. 5.3.2). The Eu^{3+} ion shows in YVO_4 relatively-red emission lines ($^5D_0 - {}^7F_2$ emission transitions at 614 and 619 nm), and thermal quenching occurs only above 300°C [30]. The thermal quenching is due to the fact that the VO_4^{3-} group shows pronounced thermal quenching at those temperatures. This was studied in a diluted system with the same crystal structure, viz. $YP_{0.95}V_{0.05}O_4$ [31].

A problem with this phosphor is that without precautions it usually contains a small amount of unreacted V_2O_5 which lowers the light output. For application in high-pressure mercury vapor lamps this phosphor is usually prepared with an excess of boric acid. The material has then a white body color and, in addition, the particle size can be controlled. The boron is not built into the lattice; in some way or another the boron compound acts as a flux.

6.4.4 Phosphors with Two-Photon Emission

All phosphors discussed up till now yield at best one visible photon for every absorbed ultraviolet photon. Taking the energy of short-wavelength radiation as about $40\,000$ cm^{-1}, and that of "average" visible light (500 nm) as $20\,000$ cm^{-1}, we loose about half of the energy. This energy is given up as heat to the host lattice in the several relaxation processes (Chapters 2 and 3); it is not a nonradiative transition which competes with the radiative one (Chapter 4).

A much higher energy efficiency can be obtained if the absorbed ultraviolet photon would split into two visible photons. Such a process would, in principle, have a quantum efficiency of 200%.

Two-photon emission has been reported for Pr^{3+} $(4f^2)$ in YF_3 [32,33] with a quantum efficiency of about 145%. The energy level scheme is given in Fig. 6.23. The Pr^{3+} ion decays in two steps as indicated. The $^1S_0$ level is at $46\,500$ cm^{-1}, and the $^3H_4 \rightarrow {}^1S_0$ transition parity forbidden. Excitation in a lamp has to occur in the $4f^{5d}$ configuration of Pr^{3+} which is above the $^1S_0$ level. Therefore the low-pressure mercury discharge cannot be applied and another solution has to be found.

Although two-photon emission would greatly enhance the efficiency of luminescent lamps, this concept has, up till now, found no practical realisation for lack of suitable materials and ultraviolet sources.

6.5 Outlook

It will be clear from this chapter that the introduction of rare-earth activated luminescent materials has drastically changed the situation. Apart from the cheaper halophosphate phosphors, we have now available a family of rare-earth activated phosphors which make luminescent lighting ideal. Not only the light output is high, but also the color rendering is excellent. For even better color rendering, the special deluxe lamps give a very good solution, although the light output is lower. The maintenance of these tricolor lamps is also very good. It is not realistic to anticipate important breakthroughs in this field.

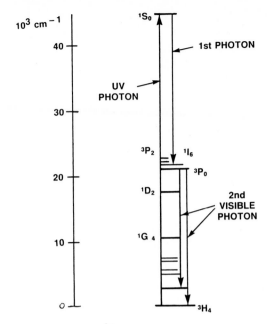

Fig. 6.23. Two-photon emission in Pr^{3+}

It is therefore to be expected that the efforts of the lamp industry will go into the following directions:

– reduction of phosphor cost
– recovery of lamps in connection with (future) environmental requirements.

As a cheaper green the composition $Sr_3Gd_2Si_6O_{18} : Pb^{2+}, Mn^{2+}$ has been suggested. However, this phosphor suffers from radiation damage [6]. A fundamental study has been devoted to the possibility of a cheaper red phosphor by activating calcium and zirconium compounds with Eu^{3+} [34,35]. In these hosts the Eu^{3+} ion carries an effective charge: $Eu_{Ca}^{\bullet}$ or Eu_{Zr}'. The positive effective charge is detrimental for the quantum efficiency of the Eu^{3+} emission upon charge-transfer excitation (q_{CT}), the negative effective charge is favorable. A qualitative model based on the configurational coordinate diagram has been proposed in order to explain these results. If the Eu^{3+} ion has a positive effective charge, the ground state shrinks more than the excited charge-transfer state, so that the value of q_{CT} decreases relative to that of $Y_2O_3 : Eu^{3+}$ (see Fig. 6.24). The only way to overcome this effect is to create very stiff surroundings around the Eu^{3+} ion. In this way q_{CT} values higher than 60% have been obtained.

For the negative effective charge more promising effects are to be expected and these have in fact been observed. Some Eu^{3+} ions in $BaZrO_3 : Eu^{3+}$ have a q_{CT} close to 100%. However, other Eu^{3+} ions have a much lower q_{CT}. Although it has become clear that cheaper lamp phosphors are not easy to find, this is not impossssible in itself.

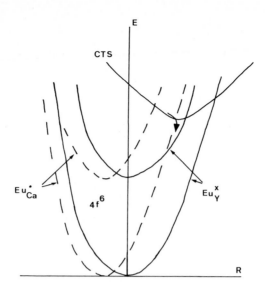

Fig. 6.24. Schematic configurational coordinate diagram for Eu^{3+} in oxides. Continuous lines relate to Eu^{3+} in an yttrium compound (e.g. Eu^x_Y in $Y_2O_3 : Eu^{3+}$). Broken lines relate to Eu^{3+} in a calcium compound (e.g. $Eu^{\bullet}_{Ca}$ in $CaSO_4 : Eu^{3+}$). The charge-transfer state is indicated by *CTS*. This state populates the emitting level in the case of Eu^x_Y, but the ground level in the case of $Eu^{\bullet}_{Ca}$. See also text

Another trend in luminescent lamp research is the extension of the measurement of excitation spectra into the vacuum ultraviolet (see for example Ref. [36]). This has nowadays become feasible by using the possibilities of a synchrotron facility. High-energy excitation can also result in quantum efficiencies above 1. For example, $Y_2O_3 : Eu^{3+}$ shows a quantum efficiency of about 100% for 6 eV excitation, and 200% for 17 eV. This value increases to 240% for 23 eV [36]. These high values were ascribed to an interband Auger process. This implies the following: upon excitation, an electron from the valence band is brought high into the conduction band. The excess energy is used to excite another electron from the valence band into the conduction band, so that one exciting photon gives two (or more) electron-hole pairs which can recombine on the luminescent center. A knowledge of the excitation spectrum over a wide energy range is useful if one looks for other possibilities than mercury for the vapor discharge in the lamp.

For more specialized applications it might be that new phosphors are still to be found. Here we draw attention to a complete different application of photoluminescence proposed recently, viz. the use of ultraviolet headlights on cars in combination with luminescent marking of all road signs [37]. Experiments have shown that ultraviolet light can double the field of vision when compared with normal dipped lights. The introduction of this system seems to be simple and promises a higher road safety and even considerable cost savings. Other applications of this method were mentioned in the original report [37].

References

1. Ouweltjes JL (1965) Modern Materials, vol 5, Academic, New York, p 161
2. Butler KH (1980) Fluorescent lamp phosphors, The Pennsylvania State University Press, University Park
3. Smets BMJ (1987) Mat Chem Phys 16:283; (1991) In: DiBartolo B (ed) Advances in nonradiative processes in solids, Plenum, New York, p 353; (1992) In: DiBartolo B (ed) Optical properties of excited states in solids, Plenum, New York, p 349
4. See e.g. West AR (1984) Solid state chemistry and its applications, Wiley, New York, Chapter 2
5. Maestro P, Huguenin D, Seigneurin A, Deneuve F, Le Lann P, Berar JF (1992) J Electrochem Soc 139:1479
6. Verhaar HCG, van Kemenade WMP (1992) Mat Chem Phys 31:213
7. Bailar Jr JC, Emeleus HJ, Nyholm R, Trotman-Dickenson AF (eds) (1973) Comprehensive Inorganic Chemistry, vol 1, p 540, Pergamon, Oxford
8. See e.g. Soules TF, Bateman RL, Hewes RA, Kreidler ER (1973) Phys. Rev. B7 1657
9. Kröger FA (1973) The chemistry of imperfect crystals, 2nd edition, North-Holland, Amsterdam
10. Mishra KC, Patton RJ, Dale EA, Das TP (1987) Phys. Rev. B35 1512
11. Oomen EWJL, Smit WMA, Blasse G (1988) Mat Chem Phys 19:357
12. Jenkings HG, McKeag AH, Ranby PN (1949) J Electrochem Soc 96:1
13. Verstegen JMPJ, Radielovic D, Vrenken LE (1974) J Electrochem Soc 121:1627
14. Koedam M, Opstelten JJ (1971) Lighting Res. Technol. 3:205
15. Buijs M, Meijerink A, Blasse G (1987) J Luminescence 37:9
16. van Schaik W, Blasse G (1992) Chem Mater 4:410
17. Smets BMJ, Verlijsdonk JG (1986) Mat Res Bull 21:1305; Ronda CR, Smets BMJ (1989) J Electrochem Soc 136:570
18. Smets BMJ, Rutten J, Hoeks G, Verlijsdonk J (1989) J Electrochem Soc 136:2119
19. Saubat B, Vlasse M, Fouassier C (1980) J Solid State Chem 34:271
20. Verstegen JMPJ, Sommerdijk JL, Verriet JG (1973) J Luminescence 6:425
21. van Schaik W, Lizzo S, Smit W, Blasse G (1993) J Electrochem Soc 140:216
22. Blasse G (1988) Progress Solid State Chemistry 18:79
23. van Schaik W, Blasse G, to be published
24. Blasse G (1986) J Solid State Chem 62:207
25. Atkins PW (1990) Physical Chemistry, 4th ed Oxford University, Oxford
26. Blasse G (1992) Int Revs Phys Chem 11:71
27. Blasse G, de Korte PHM (1981) J Inorg Nucl Chem 43:1505
28. Donker H, Smit WMA, Blasse G (1989) J Electrochem Soc 136:3130
29. Levine AK, Palilla FC (1964) Appl Phys Letters 5:118
30. Wanmaker WL, ter Vrugt JW (1971) Lighting Res Techn 3:147
31. Blasse G (1968) Philips Res Repts 23:344
32. Sommerdijk JL, Bril A, de Jager AW (1974) J Luminescence 8:341
33. Piper WW, de Luca JA, Ham FS (1974) J Luminescence 8:344
34. van der Voort D, Blasse G (1991) Chem Mater 3:1041
35. Alarcon J, van der Voort D, Blasse G (1992) Mat Res Bull 27:467
36. Berkowitz JK, Olsen JA (1991) J Luminescence 50:111
37. Bergkvist L, Bringfeldt G, Fast P, Granstrom U, Ilhage B, Kallioniemi C (1990) Volvo Technology Report, p 44
38. Struck CW, Fonger WH (1970) J Luminescence 1,2:456

CHAPTER 7

Cathode-Ray Phosphors

7.1 Cathode-Ray Tubes: Principles and Display

Devices in which phosphors are excited by means of cathode rays have great practical importance: cathode-ray tubes are used for television, oscilloscopes, electron microscopes, etc. Cathode rays are a beam of fast electrons; the accelerating voltage in a television picture tube is high (> 10 kV). Figure 7.1 presents a schematic picture of such a tube. The electron beam can be deflected by a magnetic field.

In a color television tube the luminescent screen consists of a regular array of three kinds of phosphor dots: red-emitting dots, green-emitting dots and blue-emitting dots. There is an electron gun for the red dots giving a red picture, a gun for the green dots and one for the blue dots.

As we have seen in Sect. 6.2, additive mixing of primary colors in the form of blue, green and red phosphor emissions allows the production of all colors within the triangle enclosed in the chromaticity diagram (Fig. 7.2). For black-and-white television a bluish-white emission color is preferred. This can be obtained by several

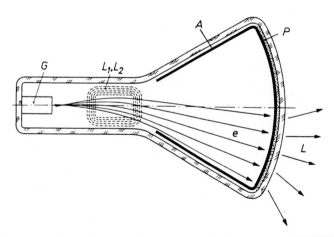

Fig. 7.1. Principle of the cathode-ray tube. Electrons (*e*) leaving the electron gun (*G*) are deflected by the systems L_1, L_2, and excite the luminescent material *P*. *A* is the anode and *L* the emitted radiation

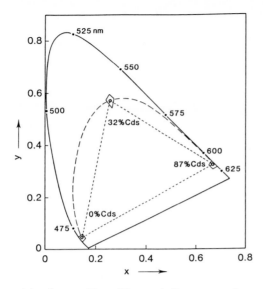

Fig. 7.2. CIE chromaticity diagram. The solid curve indicates monochromatic radiation marked by its wavelength in nm. The small quadrilaterals correspond with internationally agreed tolerance ranges. See also text

mixtures of two phosphors. In color television the problem is more complicated. The three phosphors have to be selected in such a way that the brightness is high and colors are well reproduced. Note that this is a problem different from that in lighting, where the color reproduction of illuminated objects is essential (Sect. 6.2).

Although great progress in other display techniques has been made in recent years, for example liquid crystal and electroluminescent panels, the properties of the cathode-ray tube remain up till now unsurpassed. This can be attributed to several factors. Among these are the high radiant efficiency of modern cathode-ray phosphors, their high brightness, their long life time in the tube, and the ease with which large-area uniform layers can be deposited.

Since we are dealing with a vacuum tube, the upper limit of the screen which can be reached by present-day technology is about 75 cm. Pictures with a 2 m diameter can be obtained in projection television (PTV). For each of the three colors a small (monochrome) cathode-ray tube is used. Their images are optically projected and superimposed on a projection screen using a lens system. In such a way a composite picture in full color is shown on the screen (Fig. 7.3). In order to obtain high illumination levels on the large screen, much higher current densities have to be used in PTV than in direct-view cathode-ray tubes.

Unfortunately the luminescence output is no longer linearly proportional with current density for high excitation densities (as it is for low excitation densities). Saturation occurs. Actually it is hard to find materials which satisfy all the requirements, as we will see later (see below).

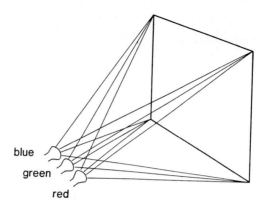

Fig. 7.3. Principle of projection television. The images of three smaller cathode-ray tubes (blue, green and red) are projected to a large screen

There are many other types of cathode-ray tubes which find application: oscillo-scope tubes, tubes with very short decay times, radar tubes, tubes with high-resolution screens, and so on. For these tubes and the phosphors they require, we refer to a recent review paper [1].

7.2 Preparation of Cathode-Ray Phosphors

Much of what has been said in Sect. 6.3 on lamp phosphors, can be repeated here. Since fast electrons excite the host lattice and create mobile electrons and holes, the restrictions to the impurity level are usually even more severe. This is certainly the case for the frequently applied sulfides. In principle zinc sulfides are made by dissolving ZnO in sulfuric acid yielding an aqueous solution of $ZnSO_4$. Subsequently H_2S is bubbled through the solution, converting $ZnSO_4$ into insoluble ZnS. The shape, size and crystal quality of this precipitate depends on the reaction conditions (pH, temperature, concentration).

The raw material is then fired with a flux and the activator, and the product is sieved. Finally the flux is removed and the product milled. Suitable processes may be applied to obtain particles of the described size ($\sim 5~\mu$m).

The red-emitting $Y_2O_2S : Eu^{3+}$ is made from a mixture of the oxides and elemental sulfur which is heated in a flux consisting of sodium carbonate and alkaline phosphate. The fired product is washed with diluted HCl in order to remove Na_2S. Figure 7.4 shows a scanning electron microscope photograph of an oxysulphide phosphor.

Several methods are in use to fabricate the luminescent screen in the cathode-ray tube. For conventional applications the screen thickness is 15–30 μm, i.e. 2–4 particles deep, resulting in a weight of 3–7 mg/cm^2. The setting method has been used since the beginning of tube manufacturing and is still in use for monochrome

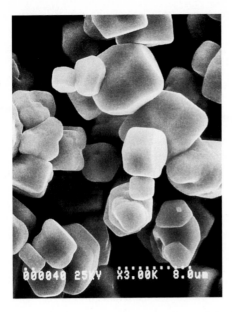

Fig. 7.4. Scanning electron microscope photograph of particles of the Y_2O_2S : Eu phosphor. (reproduced with permission from Ref. [1])

cathode-ray tubes. In order to produce color tubes, predominantly a slurry method is used. Further a dusting and an electrophoretic method can be used. For more details on screening, see Ref. [1].

7.3 Cathode-Ray Phosphors

7.3.1 Some General Remarks

The host lattices which yield the highest radiant efficiency (Sect. 4.3) for cathode-ray excitation are undoubtedly ZnS and its derivatives. For the blue-emitting ZnS : Ag values higher than 20% have been reported. We note that this fits our discussion in Sect. 4.4, where the maximum efficiency for host lattice excitation was found to occur for small values of the band gap E_g and the vibrational frequency ν_{LO} (Eqs (4.5)–(4.7)). For ZnS $E_g = 3.8$ eV and $\nu_{LO} = 350$ cm^{-1}. These values satisfy our requirements. This can be illustrated by Y_2O_3 with $E_g = 5.6$ eV and $\nu_{LO} = 550$ cm^{-1}, yielding a maximum efficiency of only 8%. Although the latter value is for red emission, this value is much lower than for ZnS.

It is in principle very simple to change the emission wavelength of the efficient blue cathode-ray phosphor ZnS : Ag$^+$, viz. by replacing part of the zinc by cadmium.

As a consequence the band gap decreases, so that the emission wavelength shifts to the red. Actually $Zn_{0.68}Cd_{0.32}S:Ag^+$ is a green-, and $Zn_{0.13}Cd_{0.87}S:Ag^+$ a red-emitting cathode-ray phosphor. The emission color is not determined by the nature of the luminescent center, but by the value of the band gap. Figure 7.2 shows the chromaticity diagram with three phosphors from the $(Zn,Cd)S:Ag^+$ family.

However, the $(Zn,Cd)S$ system has several disadvantages. In the first place, the use of cadmium has become inacceptable for environmental reasons. The red phosphor on this basis has still another large disadvantage, viz. in order to obtain red emission the larger part of the broad band emission of this phosphor is situated in the near infrared. The maximum of the emission band is close to 680 nm. This implies that the lumen equivalent of this phosphor is low (25%). For $Y_2O_2S:Eu^{3+}$ with line emission this is 55% [1].

Long ago it was predicted that color television with a satisfying brightness would only be possible with a phosphor which emits in the red by line emission around 610 nm [2]. Now we know that only the Eu^{3+} ion is able to satisfy this requirement. In fact the introduction of Eu^{3+}-activated phosphors in the color-television tube was a breakthrough: not just the red, but the total brightness increased strongly. It was also the introduction of rare-earth phosphors leading to many other improvements (e.g. in luminescent lamps) and the end of the domination of the sulfides.

7.3.2 Phosphors for Black-and-White Television [3]

In a sense this is a historical paragraph. The color preferred for black-and-white television is bluish-white. This can be realized by many combinations of two phosphors as prescribed by the chromaticity diagram. The best is $ZnS:Ag^+$ and (yellow-emitting) $Zn_{0.5}Cd_{0.5}S:Ag^+$ or $Zn_{0.9}Cd_{0.1}S:Cu,Al$. Single-component white phosphors have also been found, but none has found practical application.

7.3.3 Phosphors for Color Television

We will discuss the possible materials for the blue- , the green- , and the red-emitting phosphor in turn. For the blue the phosphor $ZnS:Ag^+$ has been continuously in use. As mentioned above, its radiant efficiency is very high and close to the theoretical limit. In Fig. 7.5 its emission spectrum is given.

The emission of $ZnS:Ag^+$ is of the donor-acceptor pair type (see Sect. 3.3.9). Silver is an acceptor in ZnS. The donor is shallow and can be aluminium or chlorine (on zinc or sulfur sites, respectively). The energy level scheme is given in Fig. 7.6.

For the green the sulphide $ZnS:Cu,Cl$ (or Al) is used. This has also an emission of the donor-acceptor pair type, but the copper acceptor levels are located at higher energy above the valence band than those of silver. This yields for copper an emission band with a maximum at 530 nm. For practical reasons it is profitable to shift the

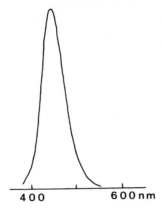

Fig. 7.5. Emission spectrum of the blue-emitting cathode-ray phosphor ZnS : Ag

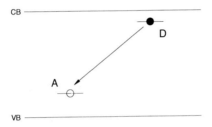

Fig. 7.6. Energy level scheme for the donor-acceptor pair emission of ZnS : Ag. The valence and conduction bands of ZnS are indicated by *VB* and *CB*, respectively. *D* is the shallow donor (aluminium or chlorine), *A* is the acceptor (silver)

emission maximum to a slightly longer wavelength. This can be done by replacing part of the zinc ($\sim 7\%$) by cadmium, or by adding a deeper acceptor such as gold. The emission color depends critically on the defect chemistry of the system [4].

For the red-emitting phosphor originally $(Zn,Cd)S : Ag$ and $Zn_3(PO_4)_2 : Mn^{2+}$ have been used. After the prediction that the red emission should consist of a narrow emission around 610 nm [2], it took another 10 years before Levine and Palilla [5] proposed $YVO_4 : Eu^{3+}$ as the red phosphor for color television tubes. This interesting material has already been discussed before (Sects. 5.3.2 and 6.4.3). A few years later it was replaced by $Y_2O_2S : Eu^{3+}$ which gave increased brightness [6]. This host lattice will reappear in the chapter on X-ray phosphors. Many years later Kano et al. proposed $Y_2(WO_4)_3 : Eu^{3+}$ because of its high lumen equivalent [7].

The great succes of the Eu^{3+} luminescence in this aspect is illustrated in Fig. 7.7. This shows the eye sensitivity curve which drops sharply in the red spectral area, and the emission spectra of a Eu^{3+}-activated phosphor and red-emitting $(Zn,Cd)S : Ag$. It is clear that the main part of the emission of the sulfide lies outside the eye sensitivity

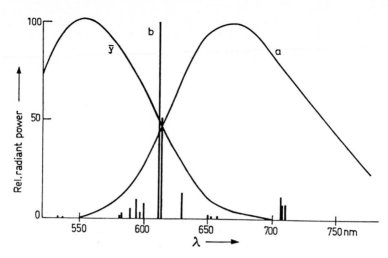

Fig. 7.7. (*a*) Emission spectrum of red-emitting (Zn,Cd)S:Ag. (*b*) emission spectrum of a red-emitting Eu^{3+} phosphor. ($\bar{y}$): eye-sensitivity curve

Table 7.1. Luminescent properties of red-emitting cathode-ray phosphors (after Ref. [1]).

Phosphor	Radiant efficiency (%)	L*	Relative brightness
$Zn_3(PO_4)_2 : Mn^{2+}$	6.7	47	39
$(Zn,Cd)S : Ag$	16	25	51
$YVO_4 : Eu^{3+}$	7	62	55
$Y_2O_3 : Eu^{3+}$	8.7	70	88
$Y_2O_2S : Eu^{3+}$	13	55	100
$Y_2(WO_4)_3 : Eu^{3+}$	4.3	81	46
611 nm radiation		100	

* Lumen equivalent relative to monochromatic light of 611 nm wavelength.

curve, so that the lumen equivalent is low. Table 7.1 gives a survey of the red-emitting cathode-ray phosphors with their properties.

It is important that the $^5D_0 - {}^7F_4$ emission of Eu^{3+} at about 700 nm is as weak as possible, since this will decrease the lumen equivalent. The high value of the lumen equivalents of $Y_2O_3 : Eu^{3+}$ and $Y_2(WO_4)_3 : Eu^{3+}$ are actually due to the low $^5D_0 - {}^7F_4$ intensity. Further, the $^5D_1$ emission of Eu^{3+} should be quenched by cross relaxation as in the lamp phosphor (see Sect. 6.4.1.4). This makes a Eu^{3+} concentration of some 3% necessary.

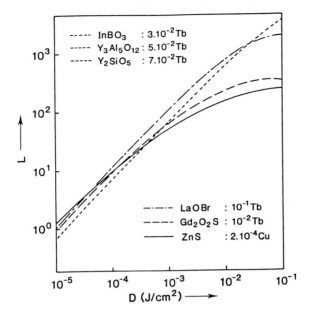

Fig. 7.8. Non-linearity of green PTV phosphors

7.3.4 Phosphors for Projection Television

As argued in Sect. 7.1 one of the problems with cathode-ray phosphors in projection television is the saturation of their light output under high excitation density. Related to this is the temperature increase of the phosphor under these circumstances: the screen temperature can rise to 100°C [8].

The nonlinear behavior of the light output was originally ascribed to ground state depletion of the activator. In sulfides, where the activator concentration is low ($\sim 0.01\%$), this is certainly important. Therefore, attention shifted to the oxidic phosphors where the activator concentrations are much higher ($\sim 1\%$). However, excited state absorption and Auger processes (Sect. 4.6) will also result in saturation effects. Detailed analysis of interactions between excited activator ions are available [9,10].

In Fig. 7.8 the light output of a number of green-emitting phosphors is plotted versus the density of cathode-ray excitation. In Fig. 7.9 their temperature dependence is given. Although the sulfide has the highest radiant efficiency at low excitation densities, viz. 20%, it shows a pronounced saturation. Note also the drop in light output upon increasing the temperature.

The Tb^{3+}-activated materials perform much better with the exception of $Gd_2O_2S:Tb^{3+}$. The bad performance of the latter seems to be due to its bad temperature dependence.

Serious candidates for application as a green phosphor in PTV tubes are $Y_2SiO_5:Tb^{3+}$, $Y_3Al_5O_{12}:Tb^{3+}$ and $Y_3(Al,Ga)_5O_{12}:Tb^{3+}$. The phosphor $InBO_3:Tb^{3+}$ is excellent from several points of view, but its decay time is very long, viz. 7.5 ms

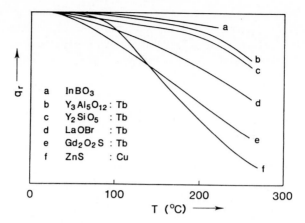

Fig. 7.9. Thermal quenching of the green emission of some cathode-ray phosphors

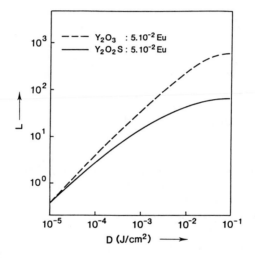

Fig. 7.10. Non-linearity of the red cathode-ray phosphors. Figs. 7.8–7.10 are reproduced with permission from Ref. [14]

[8]. This is due to the calcite structure of $InBO_3$ which places Tb^{3+} on a site with inversion symmetry which forbids the forced electric-dipole transitions (Sect. 2.3.3). An important criterion in the final phosphor selection is their degradation behavior in the tubes under high-density excitation [11].

The red phosphor in PTV tubes is usually $Y_2O_3:Eu^{3+}$, because $Y_2O_2S:Eu^{3+}$ shows saturation (see Fig. 7.10). The properties of $Y_2O_3:Eu^{3+}$ were discussed in Sect. 6.4.1.4.

An ideal blue phosphor for PTV tubes has not yet been found. Materials activated with Eu^{2+} suffer from severe degradation. The Ce^{3+} emission in $(La,Y)OBr$ and $(La,Gd)OBr$ is attractive, but the host lattice is unfavorable for screen making

[12]. The Tm^{3+} ion can give blue emission, but cross-relaxation limits the activator concentration, so that saturation occurs. In spite of its saturation at high excitation densities, the old $ZnS:Ag^+$ has not yet been surpassed under these conditions. As a consequence the screen brightness is limited by the blue-emitting phosphor. Extensive research by many laboratories has not changed this situation.

7.3.5 Other Cathode-Ray Phosphors

In the literature many other cathode-ray phosphors are known for several purposes. Two of these we will mention here, partly because they are generally known as luminescent materials, partly because their properties are also interesting from a fundamental point of view. They are $Zn_2SiO_4:Mn^{2+}$, also known by the mineral name willemite, and the family of Ce^{3+}-activated phosphors. Still more cathode-ray phosphors can be found in Ref. [1].

The green-emitting phosphor $Zn_2SiO_4:Mn^{2+}$ was also mentioned in the chapter on lamp phosphors. This might suggest that an efficient phosphor can find application everywhere. However, this is by no means true. The halophosphates, for example, are highly efficient under ultraviolet excitation (Chapter 6), but not under cathode rays. The reason for this is that efficient excitation of the activator itself is no guarantee for efficient excitation via the host lattice. However, if an activator can be efficiently excited via the host lattice, it can also be efficiently excited directly. In addition, when ultraviolet irradiation excites the host lattice, the efficiency of this excitation and cathode-ray excitation will run parallel. An illustration of the latter is formed by the sulfides and $YVO_4:Eu^{3+}$; an illustration of the former is $Y_2O_3:Eu^{3+}$ (compare the discussion in Sect. 2.1).

Willemite is used as a cathode-ray phosphor in terminal displays and oscilloscope tubes. The decay time is very long, viz. 25 ms [1]. This is mainly due to the spin- and parity- forbidden nature of the $^4T_1 \rightarrow {}^6A_1$ emission transition in the $3d^5$ configuration of the Mn^{2+} ion (Sect. 3.3.4.c), but there is also a contribution of afterglow. An even longer persistence is observed for samples to which As has been added. As a result of the As addition, electron traps are formed which trap the electrons for a certain time, so that the emission is delayed (Sect. 3.4).

Cathode-ray phosphors with such a long persistence are suitable to avoid or minimize flicker in the display. This is especially of importance when high-definition figures need to be displayed. For application in television tubes (moving pictures) or high-frequency oscilloscopes, such a long persistence is of course fatal.

The Ce^{3+}-activated cathode-ray phosphors are used in applications where a very short decay time is a requirement [13]. Since the emission is a completely allowed transition ($5d \rightarrow 4f$, Sect. 3.3.3.a), the decay time of Ce^{3+} varies between about 15 and 70 ns, depending on the emission wavelength. One application is in the beam-index tubes which generate color images by means of one electron gun [1,13]. This system could, however, never compete with the above-mentioned shadow-mask tube. The beam-index phosphor indicates the location of the electron beam. Therefore, the emission should have a very short decay time and, in order not to disturb the image, should be situated in the ultraviolet. A good choice is $Y_2Si_2O_7:Ce^{3+}$ with a radiant

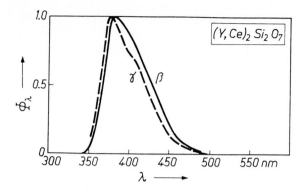

Fig. 7.11. Emission spectra of β- and γ-$Y_2Si_2O_7$: Ce^{3+} under cathode-ray excitation (after Ref. [13])

efficiency of some 7% and a decay time of 40 ns. The maximum emission wavelength is 375 nm. Figure 7.11 shows the emission spectra of the β and the γ modifications. Especially the γ modification shows clearly the double emission band character which is due to the splitting of the $4f^1$ ground configuration into the $^2F_{5/2}$ and $^2F_{7/2}$ levels (Sect. 3.3.3.a).

The other application is the flying-spot scanner. In this set up the information of slides or film can be transformed into electric signals. The electron beam excites a phosphor with short decay time. The luminescence signal scans the object point by point and the transmitted radiation is detected with a photomultiplyer. To reduce blurring of the signal, the decay time of the phosphor should be of the order of magnitude of the time the electron beam scans a picture element ($\sim$ 50 ns).

In order to transmit pictures in color, the emission of the phosphor should cover the whole visible area. For this purpose a mixture of Y_2SiO_5 : Ce^{3+} and $Y_3Al_5O_{12}$: Ce^{3+} has been used. The former emits in the blue, the latter in the green and red [13]. The reason that Ce^{3+} in $Y_3Al_5O_{12}$ emits at such long wavelengths is due to the extended crystal-field splitting of the excited Ce^{3+} ion (i.e. $5d^1$ configuration). This was considered in Sect. 3.3.3.a.

It is interesting to note that $Y_3Al_5O_{12}$: Ce^{3+} was originally developped for the flying-spot scanner tube [13]. However, today its application lies in the special de luxe lamps as discussed in Sect. 6.4.1.7. For the former application its long wave-length emission with short decay time is essential, for the latter its long wavelength absorption and emission. In both cases the low energy position of the lowest crystal-field component is essential.

7.4 Outlook

The field of cathode-ray phosphors is relatively old and has reached a high level of maturity. The quality of todays color television is a token of the success of these materials. For direct-view television there is not much more to desire as far as the luminescent materials are concerned. The combination $ZnS:Ag^+$, $ZnS:Cu^+,Al^{3+}$, and $Y_2O_2S:Eu^{3+}$ is worldwide considered to be the most suitable combination.

In the fields of projection television phosphors the situation is less satisfactory. It is clear that luminescent materials have difficulties in meeting the high requirements. At the moment the largest need is for a blue-emitting phosphor with an acceptable saturation. The bad situation is well illustrated by the blue phosphor used, viz. $ZnS:Ag^+$. Its high radiant efficiency in direct-view television tubes ($\sim 20\%$) decreases to less than 5% under the conditions of the projection-television tube. Nevertheless it has not been possible, up till now, to find an acceptable alternative.

References

1. Hase T, Kano T, Nakazawa E, Yamamoto H (1990) In: Hawkes PW (ed) Advances in electronics and electron physics, vol. 79. Academic, New York, p 271
2. Bril A, Klasens HA (1955) Philips Res. Repts 10:305; Klasens HA, Bril A (1957) Acta Electronica 2:143
3. Ouweltjes JL (1965) Modern materials, vol. 5, Academic, New York, p 161
4. Bredol M, Merikhi J, Ronda C (1992) Ber. Bunsenges. Phys. Chem. 96:1770
5. Levine AK, Palilla FC (1964) Appl. Phys. Letters 5:118
6. Royce MR, Smith AL (1968) Ext. Abstr. Electrochem. Soc. Spring Meeting 34:94; Royce MR (1968) US patent 3.418.246
7. Kano T, Kinameri K, Seki S (1982) J. Electrochem. Soc. 129:2296
8. Welker T (1991) J. Luminescence 48, 49:49
9. de Leeuw DM, 't Hooft GW (1983) J. Luminescence 28:275
10. Klaassen DBM, van Rijn TGM, Vink AT (1989) J. Electrochem. Soc. 136:2732
11. Yamamoto H, Matsukiyo H (1991) J. Luminescence 48,49:43
12. Raue R, Vink AT, Welker T (1989) Philips Techn. Rev. 44:335
13. Bril A, Blasse G, Gomes de Mesquita AH, de Poorter JA (1971) Philips Techn. Rev. 32:125
14. Smets B (1991) In: DiBartolo B (ed) Advances in nonradiative processes in solids, Plenum, New York, p 353

X-Ray Phosphors and Scintillators
(Integrating Techniques)

8.1 Introduction

The terms X-ray phosphors and scintillators are often used in an interchangeable way. Some authors use the term X-ray phosphors when the application requires a powder screen, and the term scintillator when a single crystal is required. The physical processes in the luminescence of these two types of materials is, however, in principle the same and comparable to that in cathode ray phosphors (Chapter 7).

Therefore another subdivision of the broad field of X-ray phosphors and scintillators is used here, viz. materials used in applications where integrating techniques are used (Chapter 8), and materials used in applications where counting techniques are used (Chapter 9). The integrating technique measures the light intensity under continuous excitation; it is position sensitive and yields an image; a well-known example is the case of X-ray imaging in medical diagnostics. The counting technique digests the radiation excited by a single pulse; it yields the number of exciting events; a well-known example is the use of scintillators in electromagnetic calorimeters in order to count photons, electrons or other particles.

X-ray phosphors can be defined as materials which absorb X rays and convert the absorbed energy efficiently into luminescence, in practice often ultraviolet or visible emission. In this paragraph we consider the phenomenon of X-ray absorption, and the principles of some important ways of X-ray imaging and the requirements which X-ray phosphors have to satisfy in order to be promising for potential application.

In the next paragraph several aspects of materials preparation are discussed. A complicating factor is formed by the fact that such materials are applied as powders, ceramics or single crystals, depending on the application. In Sect. 8.3, the possible materials are considered following the several applications. The final paragraph presents an outlook into the future of this complicated field of materials research.

8.1.1 X-Ray Absorption

Figure 8.1. shows a schematic picture of the X-ray absorption coefficient a vs the energy E of the X rays. When X rays interact with an atom or ion, they may remove an electron from the K shell if the energy of the X-ray quantum is equal to or larger

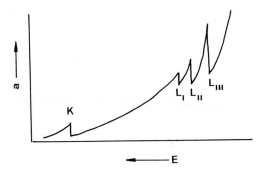

Fig. 8.1. The X-ray absorption coefficient a as a function of the energy E of the X-rays (schematic). The K and L X-ray absorption edges are indicated

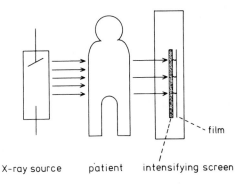

Fig. 8.2. Schematic representation of a medical radiography system based on the use of an intensifying screen

than the binding energy E_K of the K electron. This absorption yields a continuous absorption spectrum starting at E_K and extending to higher energy. This corresponds to the left-hand part of the spectrum in Fig. 8.1.

In a similar way, the weaker-bounded L electrons yield three continuous absorption spectra which start at the energies L_I, L_{II} and L_{III} (Fig. 8.2., right-hand side). Even weaker-bounded electrons (M shell, and so on) yield absorption edges at still lower energy, depending on the atomic number.

The absorption coefficient increases strongly with the atomic number. As a consequence X-ray phosphors will necessarily contain heavy elements; another way to formulate this requirement is that X-ray phosphors must be high-density materials.

8.1.2 The Conventional Intensifying Screen

After the discovery of X rays in 1895, Röntgen realized immediately that X rays are not efficiently detected by photographic film. The reason for this is the weak absorption of X rays by such a film. In this way long irradiation times would be required which results in vague pictures when the object moves (like the human body generally does). In addition the negative influence of X-ray irradiation on the human body is nowadays sufficiently known. Immediately after Röntgen's invention a search for phosphors which were able to absorb X rays and to convert the absorbed energy efficiently into light was started [1]. As early as one year later, in 1896, Pupin proposed $CaWO_4$ for this purpose This material served for some 75 years in the so-called X-ray intensifying screens, an absolute record for a luminescent material. In this way the irradiation time was reduced by about three orders of magnitude! The luminescence properties of $CaWO_4$ have already been discussed (see Chapter 1, and Sects 3.3.5 and 5.3.2).

A medical radiography system based on the use of an intensifying screen is represented in Fig. 8.2. The X-ray radiation transmitted by the patient is detected by the X-ray phosphor which is applied as a screen. The emitted luminescence is detected by photographic film. The spectral film sensitivity should coincide optimally with the spectral energy distribution of the emitted luminescence. Although the medical application of this principle is best known, other applications are also in use. An example is nondestructive materials control.

A more realistic picture of the X-ray cassette is given in Fig. 8.3. It is seen that the film is surrounded on both sides by an intensifying screen for optimal sensitivity. This figure also shows the disadvantage of the intensifying screen, viz. it causes a certain blur which impairs the definition of the image. This is due to the fact that the direction of the light emission is independent of the angle of incidence of the X-ray photon, so that the emission diverges in all directions. This is worse, since the

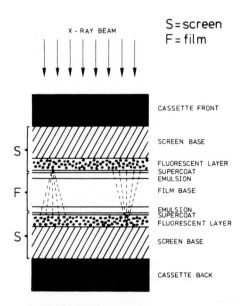

Fig. 8.3. Section through an X-ray cassette containing a double-coated film (*F*) and a pair of intensifying screens (*S*)

diffusing emission light reaches not only the film side adjacent to the emitting screen, but also that remote from the screen (see Fig. 8.3). This is known as the cross-over effect. Occasionally radiologists prefer, therefore, to make X-ray images without an intensifying screen. This is sometimes possible, for example, for an image of the hand which is a non-moving object.

Of course the packing and the size of the crystallites in the screen and the thickness of the screen are another source of unsharpness in the image. It is obvious that the smaller the crystal size, the closer the packing, and the thinner the screen layer is, the better the sharpness will be.

From this discussion it will be clear that the requirements for X-ray phosphors to be used in intensifying screens are the following: high X-ray absorption and high density, a high conversion efficiency for X-ray to light conversion, an emission spectrum which covers the film sensitivity (in practice blue or green), stability, and an acceptable cost price. The factors determining a high conversion efficiency were discussed in Sect. 4.4. Although $CaWO_4$ has been used for a long time, it did not satisfy these requirements, as we will see below. It has been replaced by rare-earth doped materials.

8.1.3 The Photostimulable Storage Phosphor Screen

About ten years ago the Japanese Fuji corporation introduced a new technique for X-ray imaging [2]. This technique is based on the use of a photostimulable storage phosphor screen. The operation of a storage phosphor is depicted schematically in Fig. 8.4. Upon irradiation electrons are promoted from the valence band to the conduction band. In a storage phosphor a number of the created free charge carriers are trapped in electron traps and hole traps. The traps are localized energy states in the bandgap due to impurities or lattice defects. If the trap depth ΔE is large compared to kT, the probability for thermal escape from the trap will be negligibly small and a metastable situation is created.

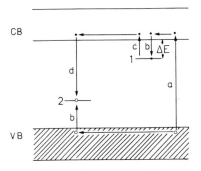

Fig. 8.4. Energy band model showing the electronic transitions in a storage phosphor: (*a*) generation of electrons and holes; (*b*) electron and hole trapping; (*c*) electron release due to stimulation; (*d*) recombination. Solid circles are electrons, open circles are holes. Center 1 presents an electron trap, center 2 a hole trap

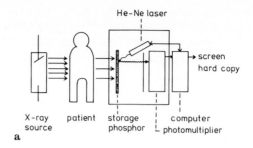

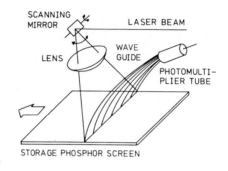

Fig. 8.5. (*a*) Schematic representation of a medical radiography system based on the use of a photostimulable storage phosphor. (*b*) The image reader in greater detail. Figures 8.2, 8.4 and 8.5 are from A. Meijerink, thesis, University Utrecht, 1990

The stored energy can be released by thermal or optical stimulation. In the case of thermal stimulation the irradiated phosphor is heated to a temperature at which the energy barrier ΔE can be overcome thermally. The trapped electron (or hole) can escape from the trap and recombine with the trapped hole (or electron). In the case of radiative recombination, luminescence is observed which is called thermally stimulated luminescence (TSL) (compare Sect. 3.5). Under optical stimulation the energy of an incident photon is used to overcome ΔE. The luminescence due to optical stimulation is called photostimulated luminescence (PSL). The phenomenon of stimulated luminescence from storage phosphors has been known since 1663 (Boyle)[2]. Storage phosphors have found a wide range of applications, e.g. as infrared detectors and in the field of dosimetry [3].

The X-ray imaging system based on a photostimulable storage phosphor is depicted in Fig. 8.5. The X-ray film is replaced by a storage phosphor screen as a primary image receptor. The transmitted X-ray photons are absorbed by the storage phosphor in the screen, where a dose-proportional amount of energy is stored. The latent image in the screen is read out by scanning the phosphor screen with a focussed

---

2 Boyle reported in the Register of the Royal Society (1663) 213 "a glimmering light from diamond by taking it to bed with me and holding it a good while upon a warm part of my naked body".

He–Ne laser. The red laser light ($\lambda = 633$ nm) stimulates recombination, resulting in photostimulated luminescence. The intensity of the photostimulated luminescence is proportional to the X-ray dose. For each spot of the laser beam on the screen the intensity of the photostimulated luminescence is measured by a photomultiplier tube and stored in a computer. The X-ray photograph in the computer can be visualized on a monitor or by making a hard copy.

This new technique for X-ray imaging offers several important advantages over conventional screen-film radiography. The response of the system is linear over at least four decades of the X-ray dose ($10^{-2} - 10^2$ mR). This wide dynamic range prevents overexposure and underexposure. The sensitivity of the system is higher due to a higher sensitivity of the photomultiplier tube to light compared to film. The higher sensitivity enables a reduction of the exposure time. Finally, the digitized images obtained with this system can be processed by a computer which offers the possibility of digital manipulation and facilitates archiving.

Apart from the high price, the main disadvantage of the digital X-ray imaging system is the lower resolution. Due to scattering of the laser beam the performance of the system is worse than the performance of the conventional screen-film system. This prohibits application in areas where a high resolution is required. The development of a translucent storage phosphor screen with minimal scattering could solve this problem.

A good storage phosphor has to meet a number of requirements:

a. The absorption coefficient for X-rays must be high. This can be achieved by using materials with a high density.
b. The amount of energy stored in the phosphor per unit X-ray dose has to be large in order to obtain high sensitivity.
c. A short decay time ($< 10 \ \mu s$) of the photostimulated emission is required for a fast readout.
d. The fading of the information stored in the phosphor has to be slow (preferably the information must still be present several hours after exposure).
e. Stimulation has to be possible in the red or near infrared. The stimulated emission has to be located between 300 and 500 nm where the photomultiplier tube has its highest sensitivity.

The photostimulable phosphor used at present in nearly all commercial digital X-ray imaging systems is $BaFBr:Eu^{2+}$. A physical mechanism for the PSL in Eu^{2+}-activated barium fluorohalides has been proposed by Takahashi et al. [4]. Optical, EPR and photoconductivity studies indicate that upon X-ray irradiation some of the holes are trapped by Eu^{2+} ions, giving Eu^{3+}, and some of the electrons in halide vacancies, giving F centers. Illumination in the F-center absorption band stimulates recombination of the trapped electron with the hole trapped on Eu^{2+}, resulting in Eu^{2+} in the excited $4f^65d$ state. The Eu^{2+} ion returns to the ground state radiatively and the characteristic Eu^{2+} emission around 390 nm is observed.

It was soon realised that the simple scheme of Fig. 8.4 is a simplification. Studies on the recombination mechanism suggest that electron-hole recombination in $BaFBr:Eu^{2+}$ does not take place via an electron entering the conduction band, but via tunnelling. The evidence for this originates from different sources [5]:

a. the decay of the photostimulated luminescence under continuous stimulation in the temperature region 100–300 K is found to be temperature independent
b. the increase of the photostimulated luminescence intensity with increasing X-ray dose appears to be linear which points to a tunnelling process
c. analysis of the thermally stimulated luminescence points also to tunnelling.

An important contribution to the theoretical model formation originates from the use of vacuum-ultraviolet radiation from a synchrotron source [6]. It was shown that photostimulable centers in $BaFBr:Eu^{2+}$ can be created by irradiation into the vacuum-ultraviolet region (> 6.7 eV), i.e. in the excitonic and interband region of the host lattice.

The model originating from this work invokes, as a first step, the relaxation of a free exciton in the neighbourhood of a lattice distortion induced by Eu^{2+}. This leads to an $e+V_K$ centre (where e stands for electron, and V_K for the V_K hole center consisting of a Br_2^- molecule on two Br^- sites). Subsequently an off-center self-trapped exciton is formed, viz. a nearest-neighbor F–H pair (where F is the well-known F center, in which an electron is trapped at an anion vacancy, and H is the H center, which can be regarded as an X_2^- molecule occupying an X^- anion site; the anion concerned here is Br^-; see also Sect. 3.3.1). This pair is assumed to be stabilized by the presence of a substitutional Eu^{2+} ion which is much smaller than Ba^{2+}. The photostimulable center is then a Eu^{2+}–F–H complex.

Photostimulation is thought to occur as follows: upon excitation, the F center induces a certain relaxation which destabilizes the H center and an excited $e + V_K$ center is formed. The excitation energy is transferred to the nearby Eu^{2+} ion and emission follows. This model also accounts for the fact that practically no Eu^{3+} has been observed after X-ray irradiation of $BaFBr:Eu^{2+}$. Note that the simple model of Fig. 8.4 requires the presence of Eu^{3+} after irradiation.

Interestingly enough, Koschnick et al. [7] have presented convincing experimental evidence from cross-relaxation spectroscopy which shows that the PSL center in $BaFBr:Eu^{2+}$ consists of a spatially correlated but undistorted F center, an O_F^x center and a Eu_{Ba}^x center. Since x indicates neutral relative to the lattice, O_F^x indicates a hole trapped on oxygen, and Eu_{Ba}^x a Eu^{2+} ion on a barium site. The oxygen is present as an impurity in the lattice. These results illustrate how difficult it is to unravel the storage mechanism in a simple compound like BaFBr. Since the latter studies were performed on crystals with low Eu concentrations, it cannot be excluded that in the powder particles in a screen the mechanism is different. In the other proposed storage phosphors the mechanism may again be different. Therefore we conclude that many details of the storage mechanism in X-ray phosphors are still unknown.

Apart from the application of X-ray storage phosphors in medical diagnostics, there are other applications foreseen [8]. Among these are the use of storage phosphors in (protein) crystallography and in the field of data storage.

8.1.4 Computed Tomography

By means of X-ray computed tomography (CT) it is possible to construct excellent cross-sectional images of the human body and head. Figure 8.6 gives some impressive examples. The prototype was introduced by Houndsfield in 1972 [9]. In such a system the X-ray tube and the detector are rigidly coupled and rotate around the patient (rotation-rotation principle) during the scan. During this rotation, a planar, fan-shaped beam of X-rays passes through a cross-sectional slice of the patients body and strikes the detector system. The principle is shown in Fig. 8.7. During the scan the tube-detector-system characteristically executes a $360°$ rotation within 1 to 2 seconds. This fan beam consists of as many individual beams as there are detector elements, nowadays normally 768 single detectors [10].

In order to be able to construct CT images, attenuation profiles of that part of the object located in the measurement field are scanned from several different viewing directions. The scanned object is continuously irradiated during the measurement. The 768 detector channels generate electrical signals corresponding to the actual beam attenuation, which are queried in rapid sequences by the detector electronics, and digitized and transmitted to the image processor.

The width of the detector elements, and the geometric arrangement of the X-ray source, the collimation and the detector determine together the spatial resolution of the system, which is at present 0.4 mm. Computed tomography detectors are operating in current mode due to a dose flux in the range of 1 R/s, which yields quanta rates up to $10^{10} s^{-1}$.

The quantitative determination of X-ray intensities with a photon energy up to 150 keV can be realized by three types of detectors: semiconductor-based detectors which call for efficient cooling, ionization chambers which have a low quantum detection efficiency ($< 50\%$), and detectors based on luminescent materials. The latter type of detector is of interest here, because it is realized by the combination of a luminescent material with a photodiode.

In the array of 768 individual detectors, each element is about one millimeter wide. The spacing between the elements is rather small, but is present to suppress cross talk by interstitials. Such a detector is shown in principle in Fig. 8.8. The X rays absorbed by the phosphor material are converted into visible light, which is detected by the coupled photodiode. The electrical signal is generated by photoelectric conversion.

The key feature of CT is not the spatial resolution but the contrast resolution, which should have an accuracy of a few parts per thousand. Very small differences in X-ray attenuation which are in general a few percent for soft tissue parts have to be detected precisely, while simultaneously CT must be able to detect the highest attenuation in the case of bone material. With modern CT installations, a few parts per thousand are recognized within a high dynamic range of up to 10^6. Therefore, the properties of a suitable CT phosphor have to satisfy the following requirements [11]:

a) The absorption coefficient should be high. This can be realized by looking for compounds which contain elements of a high atomic number, usually > 50, and a high density, > 4 g cm^{-3}.

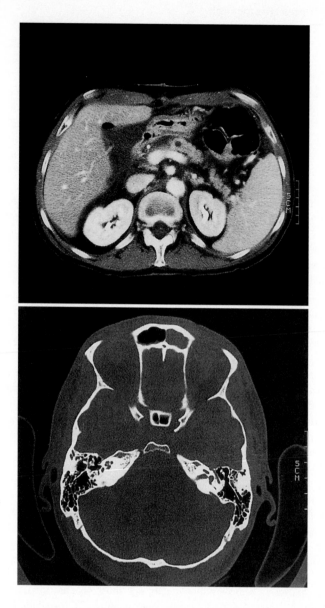

Fig. 8.6. Computed tomography images of the abdomen (top) and of the head (base of the skull) (bottom)

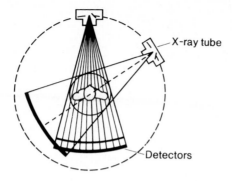

Fig. 8.7. Scanning system of X-ray CT with fan-beam design and rotation principle

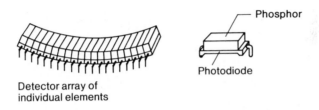

Fig. 8.8. Solid state detector scanner for CT

b) The light output should be high as a result of a high X-ray conversion efficiency and of the interior optical quality. Because the luminescent material is coupled to a Si photodiode, the emission wavelength should be close to the maximum of the sensitivity of the diode in order to ensure high signal conversion.

c) The termination of the emission after X-ray excitation should be fast, i.e. the decay time of the luminescent center has to be short, normally less than 100 microseconds.

d) The afterglow level (Sect. 3.4) should be low. If the afterglow intensities exceed specific limitations, image degeneration occurs due to memory effects. Such limitations are given by the scanning duration of the CT imaging, and should be smaller than 0.01% after 3 milliseconds.

e) Radiation damage effects should be negligibly small. The primary effect of the radiation damage is degradation in transparency rather than in luminescence.

f) Last but not least the suitable phosphor has to satisfy a number of technical requirements, concerning for example, toxicity, chemical stability, reproducibility and machining possibilities.

8.2 Preparation of X-Ray Phosphors

8.2.1 Powder Screens

Much of what has been said before about the preparation of lamp phosphors (Sect. 6.3) and cathode-ray phosphors (Sect. 7.2), is also true for X-ray phosphors from which screens are produced. Brixner [1] has described several preparation procedures. Here we give some examples.

$CaWO_4$ can be readily made via the reaction $Na_2WO_4 + CaCl_2 \rightarrow CaWO_4 + 2NaCl$. The sodium chloride formed acts as a flux to produce nice polyhedral particles of about 0.2–0.3 m^2/g surface area and an average particle size of 5–10 μ. A picture was given before (Fig. 1.7). This is a desirable particle size range, since smaller particles will lose emission intensity via internal scattering and larger particles begin to cause difficulty in making smooth thin screens. The morphology of the phosphor particles is of extreme importance. Ideally, one would like to have nearly spherical, polyhedral particles with some distribution in size. This guarantees optimal packing.

$BaFCl : Eu^{2+}$ is made from BaF_2 and $BaCl_2$ with a europium dopant via solid state reaction. Because of the layered crystal structure (see Fig. 8.9), the morphology of BaFCl crystals is plate-like and very anisotropic. Such plates pack poorly in a screen and there is a tendency for light piping towards the sides of the plates. Spray drying significantly improves the morphology. A similar problem occurs with LaOBr with the same crystal structure.

The oxysulfides Ln_2O_2S (Ln = lanthanide) (see also Sect. 7.3) are also used as X-ray phosphors. The phosphor crystallizes in nearly perfect polyhedra. Figure 8.10 gives an example for $Gd_2O_2S : Tb^{3+}$.

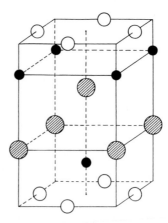

Fig. 8.9. The crystal structure of BaFCl and BaFBr. Black ions are Ba^{2+}, open ions are F^-, and hatched ions are $Cl^-(Br^-)$

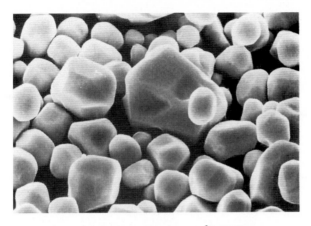

Fig. 8.10. SEM picture of the X-ray phosphor $Gd_2O_2S:Tb^{3+}$ (2500x)

8.2.2 Ceramic Plates

Within the last few years a new class of scintillators based on ceramic materials has been developed. Such ceramic scintillators are polycrystalline, inorganic, nonmetallic solids. These ceramic plates seem to be very promising for computed tomography.

The preparational route for these luminescent ceramics follows the known ceramic technologies [12,13]. They are generated by thermally induced diffusion densification of powder aggregates below their melting temperature, which is referred to as sintering. As a consequence of this preparational route the properties of such ceramic scintillators are not only determined by the host lattice and the dopants (like the luminescent activator), but additionally by the processing technique.

In contrast to conventional phosphor powders, ceramic powder synthesis aims at the generation of powders with highly sinteractive surfaces with particle sizes down to the submicron range and specific surface areas of up to 50 m^2/g. In addition, homogeneous doping on a molecular scale is of substantial importance.

For powder synthesis a large variety of methods is available, for example mixed-oxide solution precipitation or emulsion precipitation. Figure 8.11 shows as an example a $(Y,Gd)_2O_3$ powder prepared by three different precipitation techniques. The powders obtained are different in particle morphology as well as in particle size. Powders prepared by the citrate and the oxalate precipitation techniques are strongly agglomerated [14]. From powders derived in this way, compacts are formed. These powder compacts demonstrate total luminescent properties, but have still low light output due to the high porosity of about 50 vol.%.

This porosity can be reduced by a sintering process at elevated temperature. To achieve complete densification, specific sintering techniques such as vacuum sintering, hot pressing or isostatic gas pressure sintering are necessary. Figure 8.12 gives an example of a dense ceramic specimen with a homogeneous structure [15].

Fig. 8.11. Micrograph of $(Y,Gd)_2O_3$ powders prepared by emulsion precipitation, citrate precipitation and oxalate precipitation (from left to right and bottom).

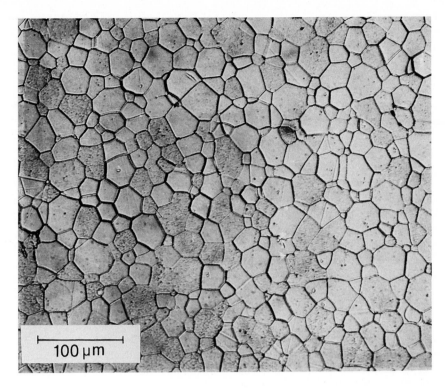

Fig. 8.12. Microstructure of a $(Y,Gd)_2O_3$ ceramic obtained by vacuum sintering at 1850°C for 2 h

8.2.3 Single Crystals

X-ray phosphors can also be single crystalline. The growth of single crystals for special applications is dealt with in Sect. 9.4.

8.3 Materials

8.3.1 X-Ray Phosphors for Conventional Intensifying Screens

The historical role of $CaWO_4$ has already been sketched above. Surprisingly enough, the successful $CaWO_4$ does not satisfy too well the requirements for X-ray phosphors as formulated above. Its X-ray absorption is relatively low, since only one out of the six constituting atoms shows strong X-ray absorption in the range 30–80 keV which is of medical importance, viz. tungsten. Also its density (6.06 g cm^{-3}) is not very high. Its broad band emission (Fig. 8.13) is hard to use completely: a blue-sensitive film

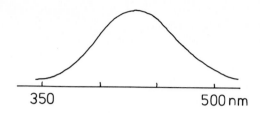

Fig. 8.13. The emission spectrum of CaWO$_4$ under X-ray excitation

does not utilize the green part of the emission. The broadness of this emission band, due to the tungstate group, was discussed in Sect. 3.3.5, and is due to strong coupling with the vibrations. On the other hand, it is an inexpensive and stable material. Most detrimental, however, is the poor X-ray-to-light conversion efficiency of only 6% (see Table 4.5). The maximum efficiency which can be calculated as described in Sect. 4.4 is found to be some 8%, so that there is not much hope of obtaining samples with higher efficiencies than obtained up till now.

Still another problem with CaWO$_4$ is the existence of a severe afterglow (see Sect. 3.4.). X-ray screens with a severe afterglow (or a long persistence) of the emission cause the appearance of a ghost image on a subsequent exposure. The afterglow of CaWO$_4$ can be reduced considerably by adding NaHSO$_4$ to the mixture to be fired. The other drawbacks of CaWO$_4$ are intrinsic, and for this reason, research for better X-ray phosphors has continued.

The first commercial rare-earth X-ray phosphor was BaFCl:Eu^{2+} proposed by the Du Pont company [1]. This material has a higher X-ray absorption and a better conversion efficiency than CaWO$_4$. However, its density is lower (4.56 g. cm^{-3}) and the morphology offers problems (see above).

Figure 8.14 shows the emission spectrum of BaFCl:Eu^{2+}. The band maximum is near the sensitivity peak of blue-sensitive film. The emission spectrum at 300 K consists of two parts, viz. sharp-line emission within the $4_f{}^7$ configuration of Eu^{2+} ($^6P_{7/2} \rightarrow {}^8S$), and band emission due to the interconfigurational transition $4f^65d \rightarrow 4f^7$. Obviously the lowest level of the $4f^65d$ configuration is only slightly higher in energy than the $^6P_{7/2}$ level (see Sect. 3.3.3b).

The emission band of BaFCl:Eu^{2+} is narrower than that of CaWO$_4$ (compare Figs. 8.13 and 8.14). This is due to the fact that the former emission belongs to the indermediate coupling regime, and the latter to the strong coupling regime (see Sect. 3.2). The importance of this difference for the fit to the film sensitivity will be clear.

A better X-ray phosphor with related characteristics is LaOBr:Tm^{3+}. This host has the same crystal structure as BaFCl (Fig. 8.9). However, the density of LaOBr is considerably higher: 6.13 g. cm^{-3}. Rabatin has contributed extensive work on phosphors based on this host lattice. It has been described in several patents [1]. Its emission spectrum is given in Fig. 8.15. It consists of intraconfigurational line transitions of Tm^{3+} ($4f^{12}$) in the near ultraviolet and blue spectral region (see Sect. 3.3.2). This phosphor is nowadays a commercial product because of its superb properties.

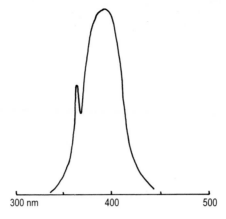

Fig. 8.14. The emission spectrum of BaFCl:Eu^{2+} under X-ray excitation. The peak at shorter wavelengths is due to the $^6P_{7/2} \rightarrow {}^8S$ transition

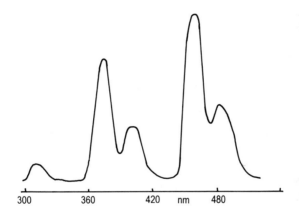

Fig. 8.15. The emission spectrum of LaOBr:Tm^{3+} under X-ray excitation

A green-emitting X-ray phosphor of high quality is Gd$_2$O$_2$S:Tb^{3+} described for the first time by Tecotzky [16]. This host also has a layered crystal structure. Using suitable preparation procedures a very nice morphology can be obtained (Fig. 8.10). The density is high, viz. 7,34 g. cm^{-3}, and the X-ray absorption favourable. The conversion efficiency of Gd$_2$O$_2$S:Tb^{3+} is about 15%, in good agreement with the maximum efficiency for this host lattice (see Sect. 4.4), and much higher than for CaWO$_4$. The Gd$_2$O$_2$S:Tb^{3+} phosphor emits in the green, showing mainly the $^5D_4$–^7F_J line transitions of Tb^{3+} (Sect. 3.3.2). In combination with a green-sensitive film this phosphor acts as a superior X-ray phosphor.

It seemed as if the introduction of LaOBr:Tm^{3+} and Gd$_2$O$_2$S:Tb^{3+} indicated the end of the search for new X-ray phosphors. However, this was not the case.

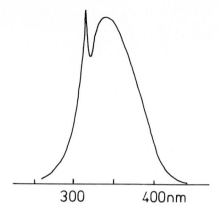

Fig. 8.16. The emission spectrum of M'–YTaO$_4$ under X-ray excitation. The peak at about 312 nm is due to a Gd^{3+} impurity

Later, other very good green-emitting X-ray phosphors were found, for example GdTaO$_4$:Tb^{3+}, Gd$_2$SiO$_5$:Tb^{3+}, and Gd$_3$Ga$_5$O$_{12}$:Tb^{3+}. Even more surprising was the invention of the tantalate-based phosphors which are physically similar to CaWO$_4$ [1,17].

The tantalate phosphors are based on M'–YTaO$_4$. This is a modification of YTaO$_4$. The other one is M–YTaO$_4$ [1] which is a distorted variant of scheelite (CaWO$_4$). In the M' modification of YTaO$_4$ the Ta^{5+} ions are in six coordination. The density is 7.55 g. cm^{-3}. To obtain particles useful for screen production a flux has to be used in the preparation (Li$_2$SO$_4$). Figure 8.16 shows the emission spectrum of M'–YTaO$_4$. The band emission is due to a charge-transfer transition in the tantalate group (Sect. 3.3.5). This phosphor is essentially an ultraviolet emitter. The replacement of part of the tantalum by niobium shifts the emission to longer wavelengths (niobate emission). The favourable X-ray absorption of YTaO$_4$ becomes clear from Fig. 8.17. Conversion efficiencies of up to 9% have been obtained. This value is equal to the maximum efficiency to be expected for this compound (see Sect. 4.4).

Even more interesting materials are phosphors based on M'–LuTaO$_4$, which is the densest white material which is not radioactive (d = 9.75 g. cm^{-3}) (1). The X-ray absorption is also higher than for the yttrium analogue (see Fig. 8.17). However, the high price of pure Lu$_2$O$_3$ will probably prevent commercial application. In conclusion, the best X-ray phosphors for the conventional X-ray intensifying screen are at the moment LaOBr:Tm^{3+}, Gd$_2$O$_2$S:Tb^{3+}, and M'–YTaO$_4$(:Nb).

8.3.2 X-Ray Phosphors for Photostimulable Storage Screens

The most popular X-ray phosphor for this purpose is undoubtedly BaFBr:Eu^{2+}. Its luminescence properties are similar to those of the isomorphous BaFCl:Eu^{2+} (Sect. 8.3.1). As discussed in Sect. 8.1.3 the storage is due to electron trapping by anion vacancies (F-center formation), and hole trapping by a variety of centers (for

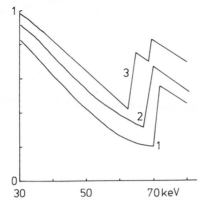

Fig. 8.17. The relative X-ray absorption of (*1*) CaWO$_4$, (*2*) YTaO$_4$ and (*3*) LuTaO$_4$ in the region of the K X-ray absorption edge (cf. Fig. 8.1)

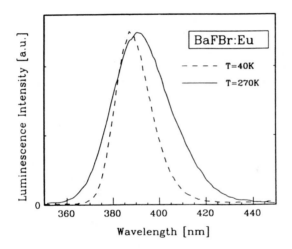

Fig. 8.18. The emission spectrum of BaFBr:Eu^{2+} at 40 and 270 K. After H.H. Rüter, thesis, University Hamburg, 1991

example an oxygen on a fluorine site, or a V$_K$ center (Sect. 3.3.1)). The photostimulable luminescent (PSL) center is thought to consist of three spatially correlated centers, viz. the electron trap, the hole trap and the luminescent center. Figures 8.18 and 8.19 show the emission spectrum and the optical stimulation spectrum of the emission of BaFBr:Eu^{2+}. This emission is due to the $4f^65d \rightarrow 4f^7$ transition on the Eu^{2+} ion. The stimulation spectrum coincides with the absorption spectrum of the F centre, indicating that the first step in the stimulation is excitation of the trapped electron.

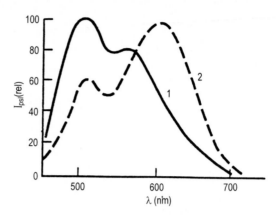

Fig. 8.19. The stimulation spectrum of BaFBr:Eu^{2+} before (*1*) and after (*2*) annealing at 550°C. The intensity of the stimulated emission (I_{psl}) is plotted versus the stimulating wavelength λ

It has been found that only the bromine F centers contribute to the photostimulability, although the X-ray irradiation creates both fluorine and bromine F centres [18]. These authors have also derived estimates of the concentrations of defect centers in a particular BaFBr:Eu^{2+} sample. Even if these values are not very reliable, they illustrate how complicated the physical mechanisms in a storage phosphor may be: 82% of the centers created by irradiation are fluorine F centers or variants thereof; these do not contribute to the photostimulable luminescence. The remaining 18% of the created centers are bromine F centers. Of these about one quarter are spatially correlated to the hole center and the Eu^{2+} ion, i.e. they yield PSL via a tunelling mechanism; the others are not correlated and need thermal activation via the conduction band in order to yield PSL. These estimated concentrations depend strongly on the history of the sample and on the Eu^{2+} concentration.

Another X-ray storage phosphor is RbBr:Tl$^+$. The luminescent center is the Tl$^+$ ($6s^2$) ion which emits by a $6s6p \rightarrow 6s^2$ transition (Sect. 3.3.7). The electron is trapped at a bromine vacancy, the hole is assumed to be trapped at a Tl$^+$ ion. The storage state can, therefore, be characterized by F + Tl^{2+}. The PSL center consists of these two centers: optical stimulation excites the F center, and the electron recombines with the hole on thallium yielding Tl$^+$ in the excited state [19]. The efficiency of the photostimulated luminescence of RbBr:Tl$^+$ decreases above 230 K due to a thermal instability of one of the trapped charge carriers.

Some other X-ray storage phosphors are the following:

– Ba$_5$SiO$_4$Br$_6$:Eu^{2+} and Ba$_5$GeO$_4$Br$_6$:Eu^{2+} [20]. Addition of a small amount of niobium improves the storage capacity and changes the storage mechanism [21]. Irradiated samples show the presence of Nb^{4+} (EPR), which is absent before irradiation. This shows that Nb^{5+} on a silicon site acts as an electron trap. Upon optical stimulation the electron becomes available for recombination with the trapped hole, the nature of which is unknown.

These materials can also reach the storage state by ultraviolet excitation. The niobate group absorbs this irradiation in a charge-transfer transition (Sect. 3.3.5). The charge-transfer state is assumed to dissociate: the electron remains at the niobium and the hole leaves the niobate group to be trapped somewhere else. The X-ray and UV excited storage states are identical.

– $Ba_3(PO_4)_2 : Eu^{2+}$. This is an efficient photoluminescent Eu^{2+} phosphor. Upon addition of a small amount of La^{3+} it obtains a large storage capacity [22]. EPR measurements on the storage state show that it contains H^0. It is assumed that the presence of La^{3+} is charge compensated by H^+ ($2Ba^{2+} \rightarrow La^{3+} + H^+$) which is available during the preparation (the phosphate source is $(NH_4)_2HPO_4$). The H^+ ion acts as an electron trap. The EPR signal disappears together with the storage state. The hole is thought to be trapped at the PO_4^{3-} group.

– $Y_2SiO_5 : Ce^{3+}$ [23]. In this material electrons are trapped by oxygen vacancies, and holes by Ce^{3+} (giving Ce^{4+}). Thermal or optical stimulation results in recombination of electron and hole on the Ce^{3+} ion yielding Ce^{3+} emission around 400 nm with a very short decay time (35 ns) (Sect. 3.3.3). The centers are spatially correlated. Codoping with Sm^{3+} changes the storage characteristics, since the electron is now (also) trapped at Sm^{3+} (giving Sm^{2+}). This is shown in Figs. 8.20 and 8.21). Photostimulation occurs by photoionization (Sect. 4.5) of the Sm^{2+} ion.

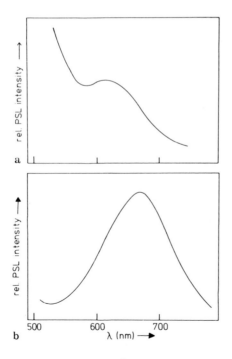

Fig. 8.20. The stimulation spectrum of the Ce^{3+} emission (400 nm) of X-ray irradiated $Y_2SiO_5 : Ce^{3+}$ (*a*) and $Y_2SiO_5 : Ce^{3+}, Sm^{3+}$ (*b*)

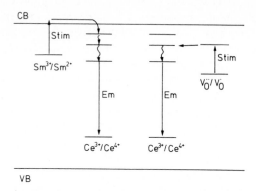

Fig. 8.21. Schematic representation of the model for the recombination in $Y_2SiO_5:Ce^{3+}$ (right) and $Y_2SiO_5:Ce^{3+}$, Sm^{3+} (left). The valence and conduction bands are given by VB and CB, respectively. *Stim* and *Em* stand for stimulation and emission, respectively

These examples show that X-ray storage phosphors have a strongly varying chemical composition which is even wider than the examples suggest. The physical mechanisms of the storage are insufficiently known. Therefore, $BaFBr:Eu^{2+}$ will not be necessarily the phosphor applied in the future.

8.3.3 X-Ray Phosphors for Computed Tomography

A detailed overview of the conventional single crystalline phosphors has been given by Farukhi [24] and Grabmaier [25]. The most important ones are $NaI:Tl$, $CsI:Tl$, $CdWO_4$, $ZnWO_4$ and $Bi_4Ge_3O_{12}$.

$NaI:Tl$ is eliminated because of serious afterglow problems, its hygroscopic behavior and its emission maximum in the blue range (415 nm) (see also Sect. 9.5.1). $ZnWO_4$ and $Bi_4Ge_3O_{12}$ have a too low light output for X-ray CT application (20% and 10% of $CsI:Tl$, respectively) (see also Sects. 9.5.2 and 9.5.3). The two remaining phosphors, $CsI:Tl$ and $CdWO_4$, have several drawbacks, as seen in Table 8.1. $CsI:Tl$ shows an excessive afterglow which cannot be influenced by growth techniques or by codoping, and its light output depends on its irradiation history. This is known as hysteresis and varies for different crystal ingots. The remaining crystal $CdWO_4$ exhibits a low afterglow, a short decay time and a sufficiently high light output (see also Sect. 9.5.2). However, the material is toxic, very brittle and has the tendency to crack parallel to its cleavage plane during machining.

A new class of phosphors based on ceramic materials has been introduced by several groups [26,27]. These materials are derivatives of the well-known rare-earth phosphor systems of oxides and oxysulfides. The most promising ceramic host lattices are $(Y,Gd)_2O_3$, Gd_2O_2S and $Gd_3Ga_5O_{12}$.

The reader will recognize these from earlier paragraphs: $Y_2O_3:Eu^{3+}$ as a lamp phosphor (Sect. 6.4.1.4) and a projection-television phosphor (Sect. 7.3.4),

Table 8.1. Characteristic data of potential phosphors for CT detectors.

	CsI : Tl	CdWO$_4$	(Y,Gd)$_2$O$_3$: Eu, Pr	Gd$_2$O$_2$ S : Pr, Ce	Gd$_3$Ga$_5$O$_{12}$: Cr, Ce
Type	Single crystal	Single crystal	Ceramic	Ceramic	Ceramic
Structure	Cubic	Monoclinic	Cubic	Hexagonal	Cubic
Density (g cm^{-3})	4.52	7.99	5.91	7.34	7.09
Attenuation coefficient 150 keV (cm^{-1})	3.21	7.93	3.40	6.86	4.36
Emission maximum (nm)	550	480	610	510	730
Light output 80 keV (rel.)	100	30	~ 67	~ 75	~ 60
Decay time (μs)	0.98	8.9	~ 1000	~ 3	140
Afterglow % after 3 ms	2	< 0.1	~ 3	$\lesssim$ 0.1	$\lesssim$ 0.1
Afterglow % after 50 ms	0.2	0.005	0.005	0.005	~ 0.01
Optical quality	Clear	Clear	Transparent	Transparent	Transparent
Chemical stability (attacked by)	Water	HCl	Conc. HCl	HCl	Conc. HCl
Mechanical behavior (20°C)	Plastic Deformable	Brittle Cleavable	Brittle High strength	Brittle High strength	Brittle High strength

Y$_2$O$_2$S : Eu^{3+} as a cathode-ray phosphor (Sect. 7.3.3) and Gd$_2$O$_2$S : Tb^{3+} as an X-ray phosphor for intensifying screens (Sect. 8.3.1), and the garnets Y$_3$Al$_5$O$_{12}$: Ce^{3+} and (Y,Gd)$_3$(Al,Ga)$_5$O$_{12}$: Tb^{3+} as a phosphor in special deluxe lamps (Sect. 6.4.1.7) and as phosphors for projection television tubes (Sect. 7.3.4) and X-ray intensifying screens (Sect. 8.3.1), respectively.

For C.T. an efficient activator for (Y,Gd)$_2$O$_3$ is the Eu^{3+} ion with its characteristic ^{5}D$_0$–^{7}F$_J$ emission (Sect. 3.3.2); its decay time is actually too long for this application. In Gd$_2$O$_2$S the Pr^{3+} ion is used as an activator with mainly ^{3}P$_0$ → ^{3}H$_J$, ^{3}F$_J$ emission and a decay time of 3 μs. In Gd$_3$Ga$_5$O$_{12}$ the Cr^{3+} ion is used. It yields broad-band (infra)red emission due to the ^{4}T$_2$ → ^{4}A$_2$ crystal-field transition (Sect. 3.3.4). This emission occurs because the crystal field at the Cr^{3+} ion in Gd$_3$Ga$_5$O$_{12}$ is relatively weak. Due to the parity selection rule (Sect. 2.1) the decay time is relatively long (140 μs). The luminescence properties of ceramic scintillators do not differ significantly from those of the related crystal or powder samples, but deviations occur in the light output depending on the different optical behaviour of ceramics.

The afterglow (Sect. 3.4), which is strongly dependent on the defect concentration in the host lattice, can be lowered by special codopants. Such an efficient codopant is praseodymium in (Y,Gd)$_2$O$_3$: Eu^{3+}, and cerium in Gd$_2$O$_2$S : Pr^{3+} as well as in Gd$_3$Ga$_5$O$_{12}$: Cr. However, it should be realized that a decrease of the afterglow level

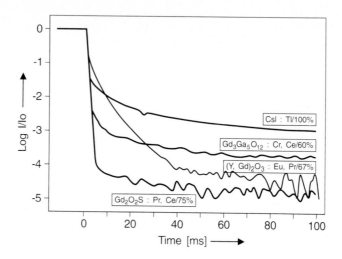

Fig. 8.22. Afterglow behavior of various scintillators for X-ray CT excitation with respect to their light output

by codoping is always correlated with a decrease of the light output. Models which try to explain how the codopants lower the afterglow level can be found in refs [28] and [29]. In summary, Fig. 8.22 shows the afterglow behaviour of various scintillators with respect to their light output, and Table 8.1 shows the properties of some ceramic phosphors compared to the single crystals CsI : Tl and CdWO$_4$.

8.4 Outlook

The search for new X-ray phosphors for conventional X-ray intensifying screens is reaching its end. A number of satisfactory materials are available, and efficiencies cannot be expected to become much higher. For specialized applications there may be some need for improvement.

Less satisfactory, but more challenging is the situation of the X-ray storage phosphors. A good material is available, viz. BaFBr : Eu^{2+}. However, the physics of the storage mechanism is incompletely understood. It is not known whether the present efficiencies are close to the theoretical limits or not. Also the development of potential applications has only just started. Therefore we may expect new results in this area in the coming years.

Computed tomography has shown considerable improvements for a couple of years. For the future the introduction of ceramic plates may be expected. Also here the introduction of new materials cannot be excluded, but seems less probable.

References

1. Brixner LH (1987) Mat. Chem. Phys. 16:253
2. Sonoda M, Takano M, Migahara J, Shibahara Y (1983) Radiology 148:833
3. McKeever SWS (1985) Thermoluminescence of solids. Cambridge University Press, Cambridge
4. Takahashi K, Miyahara J, Shibahara Y (1985) J. Electrochem. Soc. 132:1492; Iwabuchi Y, Mori N, Takahasji K, Matsuda T, Shionoya S (1994) Jap. J. Appl. Phys., 33:178
5. Meijerink A, Blasse G (1991) J. Phys. D: Appl. Physics 24:626
6. Rtter HH, von Seggern H, Reiniger R, Saile V (1990) Phys. Rev. Letters 65:2438
7. Koschnick FK, Spaeth JM, Eachus RS, McDugle WG, Nuttal RHD (1991) Phys. Rev. Letters 67:3751
8. von Seggern H (1992) Nucl. Instrum. Methods A 322:467
9. Hounsfield GN (1973) Brit. J. Radiol. 46:1016
10. Alexander J, Krumme HJ (1988) Electromedica 56:50
11. Rossner W, Grabmaier BC (1991) J. Luminescence 48,49:29
12. Kingery WD, Bowen HK, Uhlmann DR (1975) Introduction to Ceramics. Wiley, New York
13. Engineered Materials Handbook (1992) Vol. 4 Ceramics and Glasses ASM International
14. Gassner W, Rossner W, Tomande G (1991) In: Vincencini P (ed) Ceramic today-tomorrow's ceramics. Elsevier, Amsterdam, p 951
15. Grabmaier BC, Rossner W, Leppert J (1992) Phys. Stat. Sol. (a) 130: K 183
16. Tecotzky M (1968) Electrochem. Soc. Meeting, Boston, May
17. Brixner LH, Chen Hy (1983) J. Electrochem. Soc. 130:2435
18. Thoms M, von Seggern H, Winnacker A (1991) Phys. Rev. B 44:9240
19. von Seggern H, Meijerink A, Voigt T, Winnacker A (1989) J. Appl. Phys. 66:4418
20. Meijerink A, Blasse G, Struye L (1989) Mater. Chem. Phys. 21:261; Meijerink A, Blasse G (1991) J. Phys. D: Appl. Phys. 24:626
21. Schipper WJ, Leblans P, Blasse G, Chem. Mater., in press; Schipper WJ (1993) thesis, University Utrecht
22. Schipper WJ, Hamelink JJ, Langeveld EM, Blasse G (1993) J. Phys. D: Appl. Phys. 26:1487
23. Meijerink A, Schipper WJ, Blasse G (1991) J. Phys. D: Appl. Phys. 24:997
24. Farukhi MR (1982) IEEE Trans. Nucl. Sci., NS-29:1237
25. Grabmaier BC (1984) IEEE Trans. Nucl. Sci., NS-31:372
26. Greskovich CD, Cusano DA, Di Bianca FA (1986) US Pat. 4,571,312; Greskovich CD, Cusano DA, Hoffman D, Riedner RJ (1992) Am. Ceram. Soc. Bull. 71:1120
27. Yokota K, Matsuda N, Tamatani M (1988) J. Electrochem. Soc. 135:389
28. Blasse G, Grabmaier BC, Ostertag M (1993) J. Alloys Compounds 200:17
29. Grabmaier BC (1993) Proc. of the XII int. conf. defects in insulating materials. World Scientific, Singapore p 350

CHAPTER 9

X-Ray Phosphors and Scintillators (Counting Techniques)

9.1 Introduction

In Chapter 8, the excitation was exclusively by X-ray irradiation. In this chapter the stress will be on other types of ionising radiation such as γ rays and charged particles. In many cases, their energy will be higher than that of X rays. In all applications the counting of the number of ionizing events is essential. This method of radiation detection gives information on quantities such as the kind of radiation, the intensity, the energy, the time of emission, the direction and the position of the emission. Many of the applications use luminescent materials in the form of large single crystals.

The organisation of this chapter is as follows. In Sect. 9.2 the principles of the interaction between ionizing radiation and condensed matter will be dealt with. In Sect. 9.3 the principles of several applications will be discussed and the material requirements specified. The preparation of the materials will concentrate on single crystal growth and will be discussed in Sect.9.4. Finally Sect. 9.5 will give a survey of the several materials in use and Sect. 9.6 will present a future outlook. For more details than presented here, the reader is referred to the proceedings of a recent workshop [1].

9.2 The Interaction of Ionizing Radiation with Condensed Matter

There are three ways in which ionizing (electromagnetic) radiation interacts with matter, viz.

- the photoelectric effect
- the Compton effect
- pair production.

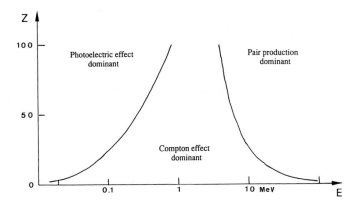

Fig. 9.1. The relative importance of the three major types of interaction of γ-rays with condensed matter. The atomic number Z is plotted linearly versus the γ-ray energy E which is plotted logarithmically (in MeV)

In the photoelectric effect, the photon is absorbed by an ion and subsequently a (photo)electron is ejected from one of the shells, usually the K shell. The photoelectron acquires an energy E_{pe} which is equal to the difference between the photon energy $h\nu$ and the binding energy E_b of the electron. This energy E_b appears in the form of X rays or Auger electrons when the vacancy in the K-shell is filled. The X rays are absorbed in a second photoelectric process and the complete energy of the incident photon is absorbed in the scintillator.

In the Compton effect, the photon interacts with an electron of an ion in the solid and transfers part of its energy to this electron. The result is a Compton scattered photon with energy $h\nu'$ $(\nu' < \nu)$ and a so-called Compton electron with energy E_c. The scattered photon may leave the scintillator or may interact with the scintillator (but at a site different from the first interaction). In the latter case the incident photon gives two light centers at different sites which makes the Compton effect undesirable for position-sensitive detection. If the scattered photon leaves the scintillator crystal, less luminescent radiation is produced than in the case of the photoelectric effect.

For photons with very high energy the process of pair production results: the photon is completely absorbed and converted into an electron-positron pair. The positron is annihilated with another electron with the emission of two photons of 0.511 MeV.

The relative importance of the three above-mentioned interaction mechanisms depends on the energy of the incident photon and on the atomic number of the absorbing ions in the scintillator. This is illustrated in Fig. 9.1.

Charged particles such as electrons, muons, or α particles lose energy through Coulomb interaction with the electrons in the solid. Two categories can be defined:

– weakly penetrating particles (low-energy electrons, protons, α particles); the rate of energy loss increases as the charge and the mass of the particle increases, whereas the scintillation yield decreases: for equal energies, a proton produces $\frac{1}{4}$

to $\frac{1}{2}$ the light of that of an electron, whereas α particles produce only $\frac{1}{10}$ of this light.

– minimum ionizing particles; they are singly charged and have low mass and high energy (fast electrons, cosmic muons). Their energy loss per unit pathlength is small.

9.3 Applications of Scintillator Crystals

Scintillator crystals are used in medical diagnostics, in industrial applications and in scientific applications [1]. A spectacular application in the latter field is the use of scintillator crystals in electromagnetic calorimeters [1,2]. These are used in high-energy physics, nuclear physics and astrophysics to count electrons and photons. The largest calorimeter was built at CERN (Geneva) in the late 1980s. It contains 12 000 $Bi_4Ge_3O_{12}$ (see Sects. 3.3 and 5.3) crystals of 24 cm in length, representing a total volume of 1.2 m^3 [2].

The energy of the radiation or the particles involved in the latter application are very high ($\sim$ GeV). As a consequence the conversion efficiency of the scintillator can be low. This efficiency is, in this field, expressed differently from the definitions given in Sect. 4.3, viz. as light yield expressed in photons per MeV. In view of the large amounts of energy collected in the calorimeter, a light yield of only 200 photons per MeV suffices. The reader should note that for emitted photons of, for example, 4 eV, this corresponds to a radiant or energy efficiency (Sect. 4.3) of 0.08%. Maximum efficiencies, estimated in Sect. 4.4, are of the order of 10–20%. It is exceptional that such a low efficiency suffices for an application.

However, in almost all other cases, a high light yield is important. The accuracy of the observation is higher if the number of observed photons, N, is larger. Energy and time resolution are proportional to $\frac{1}{\sqrt{N}}$. The basic principle of scintillation counting is that the light output of the scintillator is proportional to the energy of the incident photon [3]. In order to detect these photons, the scintillator is coupled to a photomultiplier tube in which the photons are converted into photoelectrons which are multiplied and give a pulse with an amplitude proportional to the number of photons. For a linear detector response all components should also have a linear response.

A γ-ray spectrometer should be able to discriminate between γ-rays with slightly different energy. This quality is characterized by the so-called energy resolution which depends on the light yield as indicated above.

The time resolution is defined as the ability to give accurately the moment of absorption of the photon in the scintillator. The time resolution is proportional to $\frac{1}{\sqrt{N}}$ and the decay time τ (Sect. 5.3.1). It is obvious that the moment of the absorption event is found most accurately if τ is short.

Scintillators also play an important role in medical diagnostics [4]. X-ray imaging has already been dealt with in Chapter 8. In this chapter we will mention the γ-ray camera. Radioisotopes are introduced into the body, usually in the form of chemical compounds labeled with a suitable radioactive element. A commonly used one is ^{99m}Tc

Table 9.1. Some commonly used radionuclides

Radionuclide	half life*	E_γ(keV)
^{99m}Tc	6.02 h	140.5
^{81m}Kr	13.3 s	190.7
^{123}I	13.0 h	158.9
^{67}Ga	3.26 d	93.3; 184.6; 300.3

* h: hour; s: second; d: day

Table 9.2. Some important radio isotopes in positron imaging and some examples of their application [5]

Isotope	$T_{1/2}$ (min.)	$E_\beta{}^+$ max. (MeV)	labeled substance	example of use
^{11}C	20.4	0.961	^{11}C-glucose	brain metabolism
			^{11}C-putrescine	tumor metabolism
^{13}N	9.96	1.19	^{13}NH$_3$	heart blood flow
^{15}O	2.04	1.73	^{15}O$_2$	oxygen consumption
^{124}I	4.2 days	2.13	^{124}I$^-$	thyroid studies
		1.53		

(see also Table 9.1). By measuring the radiation outside the body, a functional image can be obtained (note that the methods described in Chapter 8 were non-invasive). This emitted radiation (120–150 keV) is measured by a γ-ray camera which usually contains a scintillator crystal. Common γ-ray cameras yield a two-dimensional image of the radioactivity distribution. When the camera is rotated around the patient and/or two opposed γ-ray cameras are used, it is possible to construct a three-dimensional image (compare computed tomography, Sect. 8.1.4). This technique is called SPECT (single-photon emission computed tomography). This method does not allow to make accurate corrections for radiation attenuation in the body. Therefore, SPECT does not produce very good images, especially not of deeper-lying organs.

A different method is PET (positron emission tomography) [4,5]. It is also an in-vivo tracer technique, but uses the annihilation of positrons. Actually, one is confined to positron emitters for in-vivo studies of the distribution of elements in biologically active compounds, viz. carbon, nitrogen, oxygen. Metabolic processes can be studied in this way (see also Table 9.2).

The emitted positron cannot penetrate far into tissue; its range is only a few millimeters. After slowing down, it is annihilated with an electron. In most cases two photons (γ radiation) are emitted with an energy of 511 keV each under 180°. PET exploits the collinear emission by putting a coincidence requirement on detectors opposed to each other: an event detected simultaneously in the two detectors implies that the annihilation took place somewhere along the line between the two points of detection (Fig. 9.2). From the coincidence data, images are constructed which present the three-dimensional distribution of the radioactivity. The images are corrected for

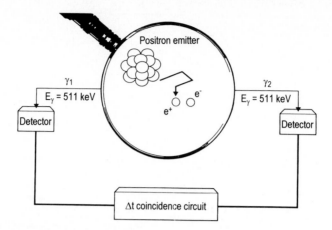

Fig. 9.2. The principle of positron emission tomography. Under the magnifying glass the positron emission has been drawn. See also text

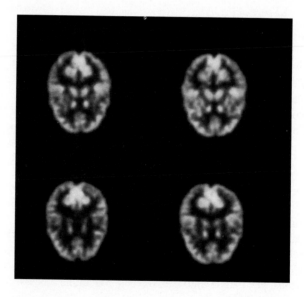

Fig. 9.3. An image of several heads obtained by positron emission tomography

radiation attenuation in the body which is not possible with SPECT. An example of a PET scan is shown in Fig. 9.3.

As mentioned above, the scintillation light can be detected using a photomultiplyer. There are also other possibilities. Among the solid-state photon detectors, silicon photodiodes have become popular. These were mentioned previously (Sect. 8.1.4,

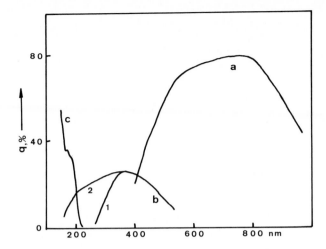

Fig. 9.4. The quantum efficiencies q(%) of the more important radiation detectors. (*a*): silicon photodiode; (*b*): photomultiplyer tube with glass (*1*) and with quartz (*2*); (*c*) solid scintillator proportional counter with gas-filled detector

computed tomography). Their sensitivity reaches a maximum for wavelengths longer than 500 nm.

A different detection unit is the solid-scintillator proportional counter [6]. It consists of a scintillator with ultraviolet emission in a multiwire chamber filled with an organic vapor. The organic molecules are ionized by the ultraviolet photons and the resulting photoelectrons are detected in the multiwire chamber. A popular molecule in this aspect is TMAE (tetrakis (dimethylamino) ethylene). Figure 9.4 presents the spectral sensitivities of the detectors mentioned.

Industrial applications form a very broad field in which imaging as well as counting techniques are applied. We can mention X-ray tomography, oil well logging, process control, security systems, container inspection, mineral processing and coal analysis [7].

Table 9.3 gives a survey of the scintillator requirements in the various applications [8,9]. As a matter of fact the atomic number of the constituents as well as the density should be high. An exception to this are scintillators for neutron detection; they should contain Li, B or Gd. Further, scintillator materials should be rugged and radiation hard. Of course they should not be hygroscopic. The emission range is, among other factors, dictated by the detector used. The light yield should be high; only in the large calorimeters used in high-energy physics in the large colliders is this requirement not essential.

Maximum possible light yields can be predicted from the maximum possible efficiency (Sect. 4.4). Some examples are given in Table 9.4 [10]. As a matter of fact the quantum efficiency of the luminescent center should be high, and competing (quenching) centers should be absent (Sect. 4.4, Refs [10,11]).

Table 9.3. Scintillator requirements in various applications (after Ref. [8])

Application	Light yield (photons/MeV)	decay time (ns)	emission (nm)
	counting techniques		
Calorimeter (high energy physics)	> 200	< 20	> 450
Calorimeter (low energy physics, nuclear physics)	hign	varies	> 300
Positron emission tomography (PET)	high	< 1	> 300*
γ-ray camera	high	less imperative	> 300
industrial applications	high	varies	> 300
	integrating techniques		
computed tomography (CT)	high	no afterglow	> 500
X-ray imaging	high	less imperative	> 350

* but < 250 nm when using a multiwire chamber

Table 9.4. Light yields of some scintillators in relation to their maximum efficiency [10]. See also text

Scintillator	light yield (ph/MeV)	η from light yield (%)	η_{max} (Sect. 4.4)
NaI : Tl	40.000	12	19
Lu_2SiO_5 : Ce	25.000	8	10
$Bi_4Ge_3O_{12}$	9.000	2	2*

* after correction for thermal quenching

Short decay times can be obtained by using luminescent ions with allowed emission transitions. In the field of inorganic materials the best examples are the $5d \rightarrow 4f$ transitions ($\tau \sim 10$ ns) (Sects. 2.3.4 and 3.3.3) and the cross luminescence ($\tau \sim 1$ ns) (Sect. 3.3.10) [11]). The afterglow is governed by the presence of traps in the host lattice as described in Sect. 3.4.

One of the most important applications of thermoluminescence (Sect. 3.5) is in the field of radiation dosimetry [12,13]. Here a material is exposed for a specific time to a radiation field. The absorbed doses ranging from 10^{-2} mGy to 10^5 Gy (1Gy (1 gray) = 100 rd (rad); 1 rad is equivalent to an absorbed energy of 0.01 J/kg) are measured by monitoring the thermoluminescence after the exposure time. Many materials display an intensity of thermoluminescence which is proportional to the amount of radiation absorbed. This led Daniels and colleagues in the early 1950s to use thermoluminescence as a means of radiation dosimetry.

The first application was in 1953 when LiF was used to measure the radiation following an atomic weapon test [12]. Later this material was used in hospital to

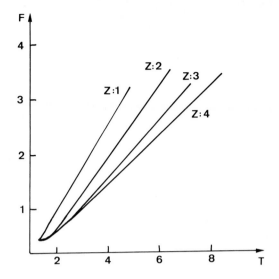

Fig. 9.5. The identification of particles in a fast component (F) vs total (T) intensity plot (BaF_2)

measure internal radiation doses received by cancer patients treated with radionuclides. Following this pioneering work, an enormous amount of work has been performed in order to extend this application. Many materials have been found (for a survey, see Ref. [12]). For example, LiF was optimized by doping with 170 mol ppm Mg and 10 mol ppm Ti.

For a further study of this application which has rather complicated aspects we refer to Ref. 12. In that book the reader will also find other applications of thermo-luminescence, e.g. age determination.

Finally we mention the possibility of particle identification by using a scintillator crystal. This method makes use of the fact that certain scintillators, for example BaF_2 and $CsI : Tl^+$, show two different emissions, one with a shorter and one with a longer decay time (see Sect. 9.5). Wisshak et al. were able to discriminate γ photons and charged hadrons using a BaF_2 scintillator [14]. A more general example, also using BaF_2, has been given by Migneco et al. [15].

The intensity ratio of the fast and the total emissions depends on the nature of the particle. This is shown in Fig. 9.5 where the fast light output is plotted vs the total light output for a BaF_2 scintillator. The contribution of the fast component decreases in the sequence γ-rays and cosmic muons, protons, deuterons, tritons and α particles. The penetration depth decreases in the same sequence, i.e. the excitation density increases. The consequence of this will be discussed in the paragraph on BaF_2 (Sect. 9.5.7).

9.4 Material Preparation (Crystal Growth)

Scintillators which are used for the detection of strongly ionizing radiation, as discussed in this chapter, are practically always applied as single crystals. The reason for this is that the scintillator should have a high optical transmission in the emission region, so that the emission can escape efficiently. This sometimes requires very large crystals without gas bubbles or precipitates, since these would result in light scattering. It is probable that certain imperfections in the crystal will also be detrimental for radiation hardness. When dopants have to be used, for example Tl^+ in NaI, the dopant will never have a uniform distribution over the whole crystal volume.

All commercially available scintillator crystals are grown from the melt. But crystals can only be grown from the melt if the compound from which a single crystal is desired has a congruent melting point, does not decompose before melting, and has no phase transition between the melting point and room temperature.

Two of the frequently applied melt growth methods are shortly discussed, viz. the Bridgman-Stockbarger method and the Czochralski method. In the Bridgman-Stockbarger technique both crystal and melt are confined within a solid container such that the three-phase boundary is between crystal, melt and container material. This technique can be divided into those in which a crystal-melt interface is propagated vertically or horizontally, those in which the whole charge is melted initially and subsequently crystallized progressively, and those in which a molten zone is established and traversed along an ingot.

Figure 9.6 shows a schematic view of the vertical directional solidification. The resistance-heated furnace is composed of several separate heating zones (in Fig. 9.6 only two zones are shown), the temperature of which can be programmed and controlled separately. The cylindrical ampoules containing the charge are supported or the furnace can be lowered during the crystal growth process.

With this methods growth rates of about one millimeter per hour can be maintained. Alkali halide crystals ranging up to 30 inches (≈ 75 cm) in diameter and half a ton of mass can be grown. Such crystals are not truly single, but rather contain five to ten components as delineated by small-angle grain boundaries which crop out at the surface or are evident under strong illumination due to scattering by impurities on the boundaries [16].

If Tl^+-activated alkali halides are desired, thallium iodide has to be mixed with the starting material (NaI or CsI) and dissolved in the melt. The distribution coefficient of thallium in the alkali halide was measured to be about 0.2 for NaI [16] and 0.1 for CsI [17], using the ratio Tl^+ concentration in the crystal/Tl^+ concentration in the melt. During crystal growth the concentration of thallium in the melt will increase, but since thallium is volatile at the melting point of NaI (T_m: 652°C) and CsI (T_m: 623°C), some of the thallium may be lost by evaporation. This means that the Tl^+ concentration in a given crystal will not be constant over the crystal volume. Fortunately the luminescence intensity is nearly constant for Tl^+ concentrations in NaI between 0.02 –0.2 mol per cent [16] and in CsI between 0.06–0.3 mol per cent

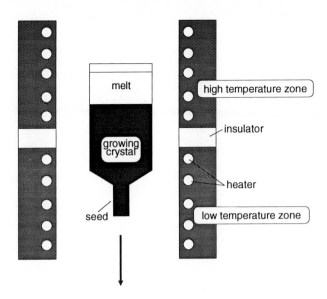

Fig. 9.6. Crystal growth according to the Bridgman-Stockbarger method

[17]. CsI:Tl crystals up to several inches in diameter can be grown out of quartz ampoules, but today Pt crucibles are used for the growth of all alkali halides.

Crystals of BaF_2 as well as CeF_3 can also be grown by the Bridgman-Stockbarger method.

The elements of the Czochralski technique are shown schematically in Fig. 9.7. The melt is contained in the crucible which is heated either by radio-frequency-induction heating or by resistance heating. The pull rod with a chuck containing the seed crystal at its lower end is positioned axially above the crucible. The seed crystal is dipped into the melt and the melt temperature adjusted until the meniscus is supported. The pull rod is rotated slowly and then lifted. By careful adjustment of the power supplied to the melt, the diameter of the crystal is controlled as it grows. Rotation rates are commonly in the range of 1–100 rpm. Pulling rates can vary from one millimeter per hour for certain oxide crystals to several tens of millimeters per hour for halide crystals. The whole assembly is enclosed within an envelope which permits control of the ambient gas. The principal advantages of the technique include the fact that the crystal remains unstrained when it cools, so that a high structural perfection can be obtained. To yield material of high and controllable purity, it is necessary to fabricate the crucible from a material which is not attacked by the molten charge.

For oxide crystals like $Bi_4Ge_3O_{12}$ (T_m: 1044°C) and $CdWO_4$ (T_m: 1272°C) Pt crucibles are normally used. For crystals with a higher melting temperature, such as Gd_2SiO_5:Ce (T_m: 1950°C), Ir crucibles are applied [18].

Oxide crystals are usually grown by the Czochralski technique, but alkali halide crystals can also be grown by this method. The perfection of these crystals is high, but the size is normally limited to 3–4 inches diameter. However, with a special

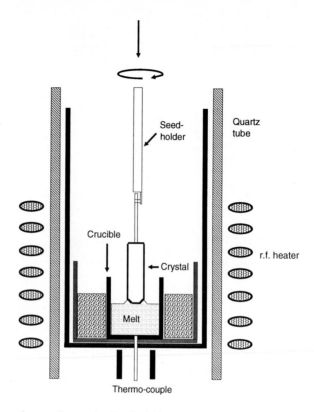

Fig. 9.7. Crystal growth according to the Czochralski method

technique, viz. simultaneous feeding of the starting material during the crystal growth process, perfect halide crystals with up to 20 inches (50 cm) diameter have been grown with a length up to 75 cm [19].

$Bi_4Ge_3O_{12}$ crystals will only be really colorless when the raw materials Bi_2O_3 and Ge_2O have a high purity (5–6N). The growth atmosphere has to be oxygen, otherwise the Pt crucible will be attacked.

The tendency of $Bi_4Ge_3O_{12}$ crystals to grow with a "core" can be partly prevented by application of a higher rotation rate up to 100 rpm. But this inconvenience can be avoided using other growth techniques. Indeed the horizontal Bridgman-Stockbarger method has quite recently become popular for $Bi_4Ge_3O_{12}$ growth, especially in China [20]. The details as to how this technique is applied are not known, but it must be successful. High quality crystals of dimensions $30 \times 30 \times 240$ mm^3 have been obtained, and are, for example applied in electromagnetic calorimeters. In order to grow large stoichiometric single crystals of CdWO$_4$ with a diameter of 3 inches, the Czochralski technique has to be slightly modified to prevent the evaporation of cadmium at the melting point of CdWO$_4$.

Table 9.5. Some properties of scintillators based on alkali halides [4,9,26]

Property	NaI : Tl$^+$	CsI : Tl$^+$	CsI : Na	CsI
density (g cm^{-3})	3.67	4.51	4.51	4.51
emission maximum (nm)	415	560	420	315
light yield (photons MeV^{-1})	40.000	55.000	42.000	2.000
decay time (ns)	230	1000	630	16
afterglow (% after 6 ms)	0.3-5	0.5-5	0.5-5	–
stability	hygro-scopic	hygro-scopic	hygro-scopic	hygro-scopic
mechanical behavior	brittle	deform-able	deform-able	deform-able

The large anisotropy of the thermal expansion along the [010] axis of Gd$_2$SiO$_5$: Ce makes the growth of single crystals rather difficult. Cracks due to cooling down from about 1950°C to room temperature occur because of large residual stresses in the grown crystal. But up to now crystals up to 2 inches in diameter have been obtained without cracks. The topics of crystal growth have been treated in many review papers and books [21–24]. As an illustration of what can be achieved today, Fig. 9.8 shows several Bi$_4$Ge$_3$O$_{12}$ crystals. Two freshly-grown ingots without cores are shown, together with several machined crystals.

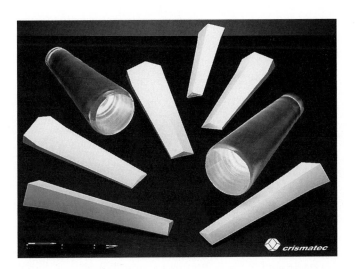

Fig. 9.8. Crystals of Bi$_4$Ge$_3$O$_{12}$ grown by the Czochralski method. Two freshly grown ingots as well as several machined crystals are shown. The authors are grateful to the Crismatex Company who made this photograph available

9.5 Scintillator Materials

9.5.1 Alkali Halides

Two of the alkali halides have been used as a scintillator material, viz. NaI and CsI, both doped with Tl^+. Table 9.5 summarizes some of their properties. Also included are CsI:Na and undoped CsI. The emission spectra of the Tl^+-doped crystals are given in Fig. 9.9.

These materials have a very high light yield (except for undoped CsI). For NaI:Tl^+, for example, the radiant efficiency calculated from the light yield is about $\frac{2}{3}$ of the maximum possible efficiency (see Table 9.4). The low light yield of CsI is, certainly for a part, due to thermal quenching [26]. For certain applications the decay time of these scintillators (< 1 μs) is acceptable. Unfortunately the afterglow is considerable and the stability poor. It depends on the application whether these scintillators can be applied or not (see Table 9.3). NaI:Tl^+ is probably the most extensively used scintillator.

The emission of the Tl^+-doped alkali halides is due to the $^3P_1-^1S_0$ transition on the Tl^+ ion (see Sect. 3.3.7). It is usually assumed that the afterglow is due to hole trapping in the host lattice (trapped exciton, see Sect. 3.3.1), whereas the electron is trapped by the activator. In CsI:Na the emission is due to an exciton bound to a Na^+ ion, in CsI to self-trapped exciton emission.

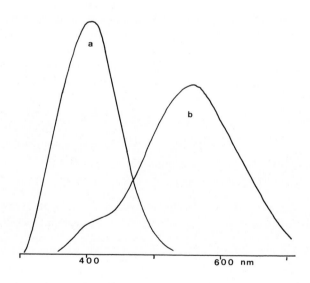

Fig. 9.9. The emission spectra of NaI:Tl (a) and CsI:Tl(b) at room temperature under X-ray excitation

Table 9.6 Some properties of the tungstate scintillators and $Bi_4Ge_3O_{12}$ [4,9,26]

Property	$ZnWO_4$	$CdWO_4$	$Bi_4Ge_3O_{12}$
density (g cm^{-3})	7.87	7.99	7.13
emission maximum (nm)	480	480	480
light yield (photons MeV^{-1})	10.000	14.000	9.000
decay time (ns)	5000	5000	300
afterglow (% after 3 ms)	< 0.1	< 0.1	0.005
stability	good	good	good
mechanical behavior	brittle	brittle	brittle

9.5.2 Tungstates

The scintillators $ZnWO_4$ and $CdWO_4$ have high densities (see Table 9.6). Their light yields are lower than for the alkali halides, but their afterglow is weak. The crystals cleave easily which makes machining difficult. $CdWO_4$ is toxic.

Their maximum efficiencies can be expected to be below 10% (Sect. 4.4). From the data in Table 9.6, $CdWO_4$ is estimated to yield 3.5%.

These tungstates show broad band emissions originating from the tungstate octahedron (in contradistinction with the scheelite $CaWO_4$, they have the wolframite crystal structure). This type of luminescence was treated in Sect. 3.3.5.

9.5.3 Bi$_4$Ge$_3$O$_{12}$ (BGO)

Some properties of BGO are summarized in Table 9.6. It is a very interesting material from the fundamental as well as from the applied point of view. It is used in calorimeters and PET scanners. The decay time is rather short, the afterglow weak, and the density high. Weber [27] has reviewed the history and properties of this scintillator. Figure 9.10 gives the luminescence spectra. The relevant optical transitions were discussed in Sects. 3.3.7 and 5.3.2.

Due to the large Stokes shift of the emission, i.e. the large relaxation in the excited state, the emission of $Bi_4Ge_3O_{12}$ is for the greater part quenched at room temperature. The calculated η_{max} is 6%, but should be reduced to 2% in view of this quenching. The actual value calculated from the data in Table 9.6 is 2%. This shows that, apart from the quenching, there are not many radiationless losses.

The compound $Bi_2Ge_3O_9$ [28] shows similar luminescence properties [29,30]. The Stokes shift is even larger than in $Bi_4Ge_3O_{12}$ (20 000 vs 17 500 cm^{-1}, respectively). As a consequence the emission intensity is quenched at 150 K, making the material unsuitable for application.

The compound $Bi_4Si_3O_{12}$ shows luminescence properties which are very similar to those of the germanium analogue [31,32].

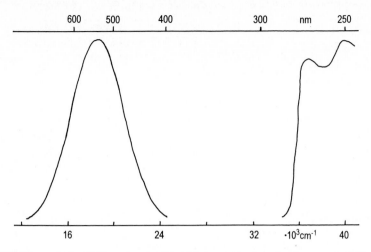

Fig. 9.10. The emission (left) and excitation (right) spectra of the luminescence of $Bi_4Ge_3O_{12}$

Table 9.7. Some properties of Ce^{3+}-activated scintillators [9,26]

Host lattice	Ce^{3+} conc. (mole %)	emission max. (nm)	light yield (photons MeV^{-1})	decay time (ns)	density (g cm^{-3})
BaF_2	0.2	310, 325	7000	60^b	4.89
LaF_3	10	290, 305	900	27	5.89
CeF_3	100	310, 340	4000	30	6.16
$YAlO_3$	0.1	350, 380	17.000	30	5.55
Gd_2SiO_5	0.5	440	9000	60^b	6.71
$Lu_2SiO_5^a$		420	25.000	40	7.4
$glass^c$	4	390	1500	70^b	2.5
$Y_3Al_5O_{12}$	0.4	550	14.000	65	5

a   C.L. Melcher, p. 415 in Ref. 1
b   and longer component
c   composition $(SiO_2)_{0.55}(MgO)_{0.24}(Al_2O_3)_{0.06}(Li_2O)_{0.06}(Ce_2O_3)_{0.04}$

9.5.4 $Gd_2SiO_5 : Ce^{3+}$ and $Lu_2SiO_5 : Ce^{3+}$

The compound Gd_2SiO_5 has a complicated crystal structure with two crystallographic sites for the rare earth ions. In recent years it has become popular as a scintillator. Crystals can be grown with the Czochralski method [19,24]. Their properties are summarized in Table 9.7. It is not hygroscopic, but it cleaves easily which can be a problem for certain applications.

The luminescence of Ce^{3+} was treated in Sects. 3.3.3 and 5.3.2. The emission transition $(5d \rightarrow 4f)$ is fully allowed, so that short decay times can be expected. This is actually observed (see Table 9.7). The variation of the decay time is mainly determined by the emission wavelength according to the relation $\tau \sim \lambda^2$ (see Sect. 3.3.3.a). The light yield of $Gd_2SiO_5 : Ce^{3+}$, although not low, is below that expected ($\eta_{observed} \sim 2.5\%$, $\eta_{max} \sim 8\%$).

Suzuki et al. [33] have reported on the ultraviolet and γ-ray excited luminescence of $Gd_2SiO_5 : Ce^{3+}$. At 11 K they were able to find luminescence from two different Ce^{3+} ions, one with an emission maximum at about 425 nm, the other with an emission maximum at about 500 nm. The respective lowest excitation bands have their maximum at 345 and 380 nm, and the respective decay times are 27 and 43 ns. The former luminescence is hardly quenched at room temperature, the intensity of the latter decreases above 200 K, and at room temperature only 20% is left. Under γ-ray excitation at room temperature the luminescence is dominated by the 425 nm emission, since the other is quenched for the greater part. Peculiarly enough, the decay shows under these conditions a long component ($\tau \sim 600$ ns) which is not observed for $Y_2SiO_5 : Ce^{3+}$ and $Lu_2SiO_5 : Ce^{3+}$.

The results of previous chapters of this book allow us to propose a simple explanation for these experimental observations. First we note that the ratio of the decay times (~ 0.65) is about equal to the squared ratio of the emission band maxima (~ 0.75) as is to be expected from $\tau \sim \lambda^2$ (see Sect. 3.3.3a).

The long decay component of $Gd_2SiO_5 : Ce^{3+}$ can be ascribed to the fact that part of the electron-hole pairs formed by γ-ray irradiation are captured by Gd^{3+} ions. The excitation energy migrates over the Gd^{3+} ions as described in Sect. 5.3.1. In this way the Ce^{3+} ions are populated in a delayed way, so that a long decay component is observed. This effect does not occur in $Y_2SiO_5 : Ce^{3+}$ or $Lu_2SiO_5 : Ce^{3+}$, since the host lattice ions do not have energy levels below the band gap energy.

The crystal structure of Gd_2SiO_5 shows that one Ce^{3+} ion is coordinated by 8 oxygen ions belonging to silicate tetrahedra and 1 oxygen which is not bounded to silicon. The latter oxygen is coordinated tetrahedrally by four rare earth ions. The other Ce^{3+} ion is coordinated by 4 oxygen ions belonging to silicate tetrahedra and 3 oxygen ions which are not. The latter Ce^{3+} ion is more strongly covalently bonded, because the oxygen ions without silicon neighbors do not have enough positive charge in their immediate surroundings to compensate their twofold negative charge [34]. Consequently, this Ce^{3+} ion has its energy levels at lower energy (see Sect. 2.2), as has also been observed for Tb^{3+} in Gd_2SiO_5 [34].

Finally the lower quenching temperature of this Ce^{3+} emission remains to be explained. It is important to note that the oxygen ions, coordinated by four rare earth ions only, form a two-dimensional network in the crystal structure of Gd_2SiO_5. The longer-wavelength emitting Ce^{3+} ions are located in this network. There is a striking structural analogy with the structure of Y_2O_3 where every oxygen is tetrahedrally coordinated by 4 yttrium ions, so that here the network is threedimensional. Actually Ce^{3+} in Y_2O_3 does not luminesce due to photoionization (see Sect. 4.5). Also in $Ca_4GdO(BO_3)_3$ the Ce^{3+} ion does not luminesce [35]. In this host lattice a similar structural network can be distinguished. Therefore we conclude that the low quenching temperature of the Ce^{3+} ion in Gd_2SiO_5 which is coordinated by three oxygen ions which do not coordinate to silicon must be explained in the same way.

The quenching of one of the two Ce^{3+} centres in Gd_2SiO_5 is at least partly responsible for the discrepancy between the observed and theoretical values of the efficiency.

The host lattice Lu_2SiO_5 has a different crystal structure which is also the structure of Y_2SiO_5. Luminescence studies do not show a large amount of quenching of the Ce^{3+} emission at room temperature [33]. It is therefore not surprising that in this structure much higher yields can be obtained than in Gd_2SiO_5. The experimental values are close to the theoretical maximum (see Table 9.4). It has been found that $Y_2SiO_5:Tb^{3+}$ yields, under X-ray excitation, also a higher efficiency than $Gd_2SiO_5:Tb^{3+}$ [34]. This suggests that the Gd_2SiO_5 structure contains in some way centers which compete with the activator ions for the capture of charge carriers. The scintillator $Lu_2SiO_5:Ce^{3+}$ seems, therefore, to have many advantages. Unfortunately, the cost price of pure Lu_2O_3 is extremely high.

9.5.5 CeF₃

Some properties of CeF_3 scintillators are gathered in Table 9.7. This material is one of the serious candidates for a new generation of high-precision electromagnetic calorimeters to be used for the new large proton collider to be built at CERN (Geneva). For that purpose one needs a total crystal volume as large as 60 m³ [2], which is nearly two orders of magnitude more than at present (1.2 m³ $Bi_4Ge_3O_{12}$). As mentioned, the relatively low light yield of CeF_3 is not detrimental for this specific application (see Table 9.3). In view of these plans, a large amount of research has already been performed on CeF_3 [36–38]. In passing, it should be mentioned that the high costs of this proposal have forced the scientists involved to look for a cheaper solution based on a reasonable compromise between cost and performance.

CeF_3 is a material with 100% activator concentration. As argued in Sect. 5.3.2, the large Stokes shift of the Ce^{3+} emission localizes the excited state, so that concentration quenching by energy migration does not occur.

In our opinion the paper by Moses et al. [36] on the scintillation mechanisms in CeF_3 is a fine example of how scintillators should be studied from a fundamental point of view. A combination of techniques was used, viz. (time-resolved) luminescence spectroscopy, ultraviolet photoelectron spectroscopy, transmission spectroscopy, and the excitation region was extended up to tens of eV by using synchrotron radiation. Further, powders as well as crystals with composition $La_{1-x}Ce_xF_3$ were investigated.

The emission depends strongly on the value of x and on the excitation energy (see also Fig. 9.11). The intrinsic Ce^{3+} emission consists of a narrow band with maxima at 284 and 300 nm. These are ascribed to transitions from the lowest level of the $5d$ configuration to the two levels of the $4f$ ground configuration ($^2F_{5/2}$, $^2F_{7/2}$) (Sect. 3.3.3). If $x > 0.1$, an additional emission band appears at longer wavelength (around 340 nm) which sometimes even dominates (see Fig. 9.11). This one is ascribed to Ce^{3+} ions close to defects.

Instructive is the emission for $x < 0.01$: it consists of the intrinsic Ce^{3+} emission band, Pr^{3+} emission lines, and a broad band extending from 250 to 500 nm which is ascribed to self-trapped exciton (STE) emission from the host. The STE consists

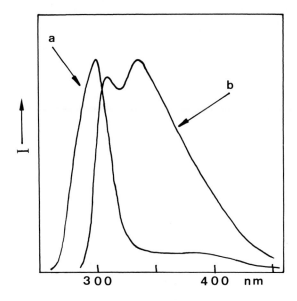

Fig. 9.11. Emission spectra of CeF$_3$. Curve *a* is for a powder under X-ray excitation, curve *b* for a crystal under γ-ray excitation. From data in Ref. [36]

of an electron bound to a V$_K$ centre which is a hole trapped by two fluorine ions forming a pseudo-molecule F$_2^-$ (Sect. 3.3.1). This shows that the lattice itself also traps the electron-hole pairs. At room temperature the STE migrates over the lattice, ending its life by radiative recombination, transfer to Ce^{3+} or to Pr^{3+} (the latter being present as an impurity), or by nonradiative recombination. This illustrates that for this composition the factor S in Eq. (4.6) is far from one if one considers the Ce^{3+} emission.

In CeF$_3$ energy transfer (Chapter 5) from intrinsic to extrinsic Ce^{3+} ions takes place [36,37]. Extrinsic Ce^{3+} ions are Ce^{3+} ions near to imperfections in the crystal lattice. Radiative as well as nonradiative transfer contributes. Actually the 340 nm emission shows a build-up which is equal to the decay of the 290 nm emission. The decay times of these emissions are about 30 and 20 ns, respectively. This agrees well with the $\tau \sim \lambda^2$ relation (Sect. 3.3.3a).

Under high-energy excitation an initial much faster decay has been observed. This phenomenon was studied by Pédrini et al. [37]. This fast component is of greater importance if there are more defects (impurities) in the lattice. However, even in very pure crystals it is present. When a high-energy particle is absorbed, the region of its relaxation has a radius of 10–100 nm. Therefore the electronic excitation is correlated in space and time. Auger relaxation in excited pairs has been proposed as a loss mechanism. The importance of this process decreases with temperature, since the excited states become more and more mobile at higher energies, so that the pair of excited ions can dissociate.

The light yield of CeF_3 is low, viz. 4000 photons MeV^{-1}. This corresponds to $\eta \sim 1\%$, whereas $\eta_{max} \sim 8\%$. This shows that the greater part of the emission is quenched. It is usually admitted that impurity rare earth ions cannot be responsible for such a loss. However, fluoride crystals will always contain a certain (low) amount of oxygen. If the presence of O^{2-} forces one of the neighboring cerium ions to become Ce^{4+} for charge compensation, a bulky quenching center is created, since intervalence charge-transfer transitions (Sect. 2.3.7) are at low energy and yield seldom emission [39]. If we add to this loss the Auger process mentioned above, it is understandable that the light yield of CeF_3 is considerably lower than is expected.

Moses et al. [36] have determined the quantum efficiency (Sect. 4.3) of the CeF_3 luminescence. For direct Ce^{3+} excitation it is high. Lower quantum efficiencies are found if the excitation starts at the F^- ion ($2p$). For 100 eV the total quantum efficiency is about 0.7. The energy efficiency is then 3%. Also this is relatively low, and the authors suggest nonradiative recombination on quenching centers in order to explain this.

9.5.6 Other Ce^{3+} Scintillators and Related Materials

The strong potential of scintillators like $Gd_2SiO_5 : Ce^{3+}$, $Lu_2SiO_5 : Ce^{3+}$ and CeF_3 have prompted a search for other Ce^{3+}-activated scintillators. Recently many new ones have been proposed [1,9,11]. Some of these have been included in Table 9.7. New ones are still appearing.

Here we mention $BaF_2 : Ce^{3+}$, $YAlO_3 : Ce^{3+}$ and $Y_3Al_5O_{12} : Ce^{3+}$ (see Table 9.7). Further there are reports on CeP_5O_{14} ($\tau \sim 30$ ns, light yield 4000 photons MeV^{-1} [11]), $LuPO_4 : Ce^{3+}$ (25, 17200 [11]), $CsGd_2F_7 : Ce^{3+}$ (30, 10000 [40]), and $GdAlO_3 : Ce^{3+}$ ($\tau \sim 1$ ns, which is very short; no light yield given [41]).

A slightly different approach is the use of Nd^{3+} as suggested by Van Eijk's group [9]. The Nd^{3+} ion has $4f^3$ configuration with a $5d \rightarrow 4f$ emission transition in the ultraviolet (~ 175 nm). Since this is an allowed transition at very high energy, the radiative decay time is even shorter than for Ce^{3+}, viz. 6 ns (in $LaF_3 : Nd^{3+}$). See also Sect. 3.3.3. The light yield is a few hundred photons per MeV. Several other host lattices have been tried, but light yields never surpass 1000 photons MeV^{-1} [42]. An important problem is formed by the absorption of Nd^{3+} emission by rare earth impurities.

9.5.7 BaF_2 (Cross Luminescence; Particle Discrimination)

The optical transitions in the luminescence of BaF_2 are discussed in Sect. 3.3.10. Two emissions are observed, a very fast cross luminescence ($\tau = 0.8$ ns, emission maxima 195 and 220 nm) and a much slower self-trapped exciton emission ($\tau = 600$ ns, maximum ~ 310 nm) which is for the greater part quenched at room temperature.

Table 9.8. Some properties of BaF$_2$

density	4.88 g cm^{-3}
maximum emission } wavelengths }	310 nm (slow) 220 and 195 nm (fast)
decay times	630 ns (slow) 0.8 ns (fast)
light yield	6.500 photons MeV^{-1} (slow) 2.500 photons MeV^{-1} (fast)

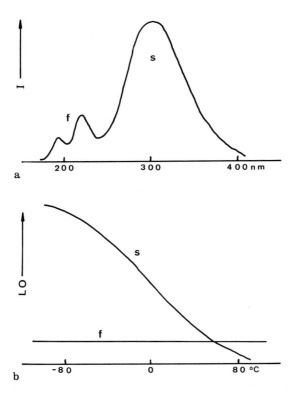

Fig. 9.12. Some data on the scintillation of BaF$_2$. (*a*): emission spectrum at room temperature under γ-ray excitation; the slow component is indicated by s, the fast component by f. (*b*): temperature dependence of the light output (*LO*) of BaF$_2$ under γ-ray excitation in a wide region around room temperature; the fast component (*f*) is temperature independent, the slow component (*s*) decreases strongly with increasing temperature

BaF$_2$ is not hygroscopic and large crystals can be grown. The slow emission component can be reduced by a factor of four by adding 1% LaF$_3$. Table 9.8 gives some of its properties, whereas Fig. 9.12 shows the emission spectrum and the temperature dependence of both components [26].

At the moment there is excellent agreement between the observed emission spectrum of BaF_2 and the spectrum obtained from an ab initio calculation on basis of a molecular cluster approach [43,44]. This confirms the spectral assignments. The relatively low light yield of the fast component is a drawback for the application. Unfortunately the slow emission component is also of an intrinsic nature and captures a greater share of the electron-hole pairs than the fast component. The whole topic has been reviewed by Van Eijk [9,45].

The BaF_2 scintillator can be nicely used for particle discrimination (see Sect. 9.3 and Fig. 9.5), because the intensity ratio of the fast and the slow components of the emission depends on the nature of the excitation. Figure 9.5 shows clearly that the heavier the exciting particles are, the less fast emission there is. The literature does not seem to contain an explanation for this effect. It should be realized that heavy particles will not penetrate deeply into the scintillator. This means that the excitation density must be very high. Since at room temperature the exciton will be more mobile than the electron-hole pair responsible for the cross luminescence, Auger interactions (Sects. 4.6 and 9.5.5) will effect the fast component more strongly than the slow component. Note an interesting wordplay here: whereas the excited state of the cross luminescence is fast in the emission process, it is slow as far as migration through the lattice is concerned (the reason for this is the strongly localized character of the hole in the Ba $5p$ core band; see Fig. 3.27). The slowly emitting exciton is fast in the lattice.

9.5.8 Other Materials with Cross Luminescence

In view of the strong interest in very fast scintillator emission, it is not surprising that many other compounds have been investigated for cross luminescence. It is essential, of course, that the cross-luminescence emission energy is smaller than the bandgap energy, since otherwise the cross luminescence cannot be emitted (see also Fig. 3.27). This is illustrated in Table 9.9 [9]. The table shows excellent agreement between prediction and observation.

Table 9.10, finally, shows some compounds for which cross luminescence has been definitely observed [9]. All decay times are of the same order of magnitude (~ 1 ns), whereas the light yields do not reach the level of 2000 photons MeV^{-1}. It is, at this time, too early to predict whether cross luminescence will have an important application or not.

9.6 Outlook

For many years scintillator research was often performed in the conviction that the physical mechanisms were unknown, that predictions were impossible, and that new materials should be found by trial and error [46]. As shown in this book and elsewhere

Table 9.9 On the possibility of cross luminescence [9]

compound	$E_c–E_{VB}$[a] (eV)	E_g[b] (eV)	predicted[c]	observed[d]
BaF_2	4.4–7.8	10.5	+	+
SrF_2	8.4–12.8	11.1	o	–/STE
CaF_2	12.5–17.3	12.6	–	–/STE
CsCl	1–5	8.3	+	+
CsBr	4–6	7.3	+	+
CsI	0–7	6.2	0	–/STE
KF	7.5–10.5	10.7	+	+
KCl	10–13	8.4	–	–/STE

a energy differences between top of core band and bottom or top of the valence band

b energy gap

c +: cross luminescence (CL) possible, –: CL impossible, o: CL doubtful

d +: CL observed, –: no CL observed, STE: exciton emission observed

Table 9.10. Scintillators with cross luminescence at 300 K [9]

compound	emission maximum (nm)	light yield (photons MeV^{-1})	τ (ns)
BaF_2	195, 220	1400	0.8
CsF	390	1400	2.9
CsCl	240, 270	900	0.9
RbF	203, 234	1700	1.3
$KMgF_3$	140–190	1400	1.5
$KCaF_3$	140–190	1400	< 2
KYF_4	140–190	1000	1.9
$LiBaF_3$	230	1400	1
$CsCaCl_3$	250, 305	1400	< 1

[10,11,47], such a conviction is not justified. The knowledge from other fields of luminescence can be very helpful. The light yield is predictable from Eq. (4.6); η_{max} is a well-known quantity in cathodoluminescence and q in photoluminescence. The transfer factor S in Eq. (4.6) is hard to predict and depends probably on the perfection and purity of the crystals. Decay times are easily predictable (see Sect. 3.3).

Modern instrumentation, in addition, offers large potential in studying fundamental processes in scintillators. In the first place the synchrotron should be mentioned which makes monochromatic excitation up to very high energies possible. Examples were mentioned above [36,37].

These developments tend to promote scintillator research into becoming "big science". Also, international cooperation is increasing (see, for example, Ref. [38]), and the multidisciplinarity of this type of research is growing. Actually, the short history of the CeF_3 scintillator sketched above illustrates all these developments very well.

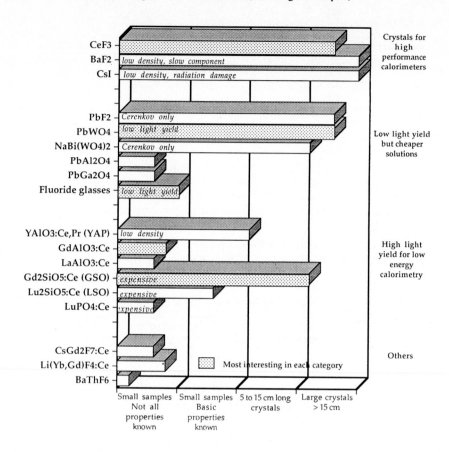

Status of some promising fast
scintillators studied by the Crystal Clear
collaboration

Fig. 9.13. An overview of scintillator research in the Crystal Clear collaboration on scintillator crystals for calorimeter application (1993). Reprinted with permission from their status report of 3rd September 1993

This does not necessarily mean that many new scintillators can be expected. The reason for this view is in the transfer factor S. In order to bring it close to 1, it is necessary to work with simple systems, and many of these have already been checked. For the same reason we do not have high expectations of amorphous scintillators. They contain "intrinsically" too many centers which will contribute to quenching. The open areas will soon be filled in using the theory available at the moment. However, to optimize a given composition to a scintillator which satisfies the requirements of the application is still a hell of a job. It requires cooperation between materials scientists of several backgrounds (crystal growth, defect chemistry, solid state physics, materials science, spectroscopy, and radiation damage). An example of such a group is the

Crystal Clear Collaboration (see authors of Ref. [38]). Figure 9.13 illustrates results from their work as summarized in the Status Report of this group of 3 September 1993. It will be interesting to see how this picture evolves in the coming years.

References

1. De Notaristefani F, Lecoq P, Schneegans M (1993) (eds), Heavy scintillators for scientific and industrial applications. Editions Frontieres, Gif-sur-Yvette
2. Lecoq P (1994) J Luminescence 60/61:948
3. Knoll GF (1987) Radiation detection and measurement, Wiley, New York
4. Grabmaier BC, p 65 in Ref. [1]; J Luminescence 60/61:967
5. Schotanus P (1988) thesis, Technical University, Delft
6. Anderson DF (1982) Phys Letters B118:230
7. Melcher CL, p 75 in Ref. [1]
8. van Eijk CWE, pp 161 and 601 in Ref. [1]
9. van Eijk CWE (1993) Nucl. Tracks Radiat. Meas. 21:5
10. Blasse G (1994) J Luminescence 60/61:930
11. Lempicki A, Wojtowicz AJ (1994) J Luminescence 60/61:942; Lempicki A, Wojtowicz AJ, Berman E (1993) Nucl Instr Methods A322:304
12. McKeever SWS (1985) Thermoluminescence of solids, Cambridge University, Cambridge; McKeever SWS, Markey BG, Lewandowski AC (1993) Nucl. Tracks Radiat. Meas. 21:57
13. Azorin J, Furetta C, Scacco A (1993) Phys Stat Sol(a) 138:9
14. Wisshak K, Guber K, Kappeler F, Krisch J, Müller H, Rupp G, Voss F (1990) Nucl. Instr. Methods A292:595
15. Migneco E, Agodi C, Alba R, Bellin G, Coniglione R, Del Zoppo A, Finocchiaro P, Maiolino C, Piattelli P, Raia G, Sapienza P (1992) Nucl. Instr. Methods A 314:31
16. Nestor OH (1983) Mat. Res. Soc. Symp. Proc. 16:77
17. Grabmaier BC (1984) IEEE Trans. Nucl. Sci. NS-31:372
18. Takagi K, Fukazawa T (1983) Appl. Phys. Lett. 42:43
19. Goriletsky VI, Nemenov VA, Protsenko VG, Radkevich AV, Eidelman LG (1980) Proc. 6th conf. on crystal growth, Moscow, p III 20
20. Chongfau H (1987) (Shanghai Institute of Ceramics, Academia Sinica, unpublished manuscript), cited as ref. 59 by Gévay G, Progress Crystal Growth Charact 15:181
21. Wilke KTh, Bohm J (1988) Kristallzüchtung. Verlag Harry Deutsch, Thun (in German)
22. Anthony AM, Collongues R (1972) In: Hagenmuller P (ed) Preparative methods in solid state chemistry. Academic, New York, p 147
23. West AR (1984) Solid state chemistry and its applications. Wiley, New York, Sect. 2.7
24. Ishii M, Kobayashi M (1991) Progress Crystal Growth Charact 23:245
25. Gévay G (1987) Progress Crystal Growth Charact 15:145
26. Schotanus P (1992) Scintillation Detectors, Saint-Gobain, Nemours
27. Weber MJ (1987) Ionizing Radiation (Japan) 14:3
28. Grabmaier BC, Haussühl S, Klüfers P (1979) Z. Krist. 149:261
29. Timmermans CWM, Boen Ho O, Blasse G (1982) Solid State Comm 42:505
30. Timmermans CWM, Blasse G (1984) J Solid State Chem 52:222
31. Blasse G (1968) Philips Res Repts 23:344
32. Ishii M, Kobayashi M, Yamaga I, p 427 in Ref. [1]
33. Suzuki H, Tombrello TA, Melcher CL, Schweitzer JS (1992) Nucl. Instrum. Methods A 320:263
34. Lammers MJJ, Blasse G (1987) J. Electrochem. Soc. 134:2068; unpublished measurements
35. Dirksen GJ, Blasse G (1993) J. Alloys Compounds 191:121
36. Moses WW, Derenzo SE, Weber MJ, Cerrina F, Ray-Chaudhuri A (1994) J Luminescence 59:89

37. Pedrini C, Moine M, Boutet D, Belsky AN, Mikhailin VV, Viselev AN, Zinin EI (1993) Chem Phys Letters 206:470
38. Anderson S, Auffray E, Aziz T, Baccaro S, Banerjee S, Bareyre P, Barone LE, Borgia B, Boutet D, Burg JP, Chemarin M, Chipaux R, Dafinei I, D'Atonasio P, De Notaristefani F, Dezillie B, Dujardin C, Dutta S, Faure JL, Fay J, Ferrère D, Francescangeli OP, Fuchs BA, Ganguli SN, Gillespie G, Goyot M, Gupta SK, Gurtu A, Heck J, Hervé A, Hillimanns H, Holdener F, Ille B, Jönsson L, Kierstead J, Krenz W, Kway W, Le Goff JM, Lebeau M, Lebrun P, Lecoq P, Lemoigne Y, Loomis G, Lubelsmeyer K, Madjar N, Majni G, El Mamouni H, Mangla S, Mares JA, Martin JP, Mattioli M, Mauger GJ, Mazumdar K, Mengucci PF, Merlo JP, Moine B, Nikl N, Pansart JP, Pedrini C, Poinsignon J, Polak K, Raghavan R, Rebourgeard P, Rinaldi DT, Rosa J, Rosowsky A, Sahuc P, Samsonov V, Sarkar S,Schegelski V, Schmitz D, Schneegans M, Seliverstov D, Stoll S, Sudhakar K, Svensson A, Tonwar SC, Topa V, Vialle JP, Vivargent M, Wallraff W, Weber MJ, Winter N, Woody C, Wuest CR, Yanovski V (1993) Nucl. Instrum. Methods A332:373
39. Blasse G (1991) Structure and Bonding 76:153
40. Dorenbos P, Visser R, van Eijk CWE, Khaidukov NM, p 355 in Ref. [1]
41. Mares JA, Pedrini C, Moine B, Blazek K, Kvapil J (1993) Chem Phys Letters 206:9
42. Visser R, Dorenbos P, van Eijk CWE, p 421 in Ref. [1]
43. Andriessen J, Dorenbos P, van Eijk CWE (1991) Molec. Phys. 74:535
44. Andriessen J, Dorenbos P, van Eijk CWE (1993) Nucl. Tracks Radiat. Meas. 21:139
45. van Eijk CWE (1994) J Luminescence 60/61:936
46. Blasse G (1991) IEEE Trans. Nucl. Science 38:30 47. Blasse G, p 85 in Ref. [1]

CHAPTER 10

Other Applications

Luminescence has lead to many more applications than those discussed in the previous four chapters. In this chapter some of these will be discussed shortly. For this purpose we selected the following topics: upconversion, the luminescent center as a probe, luminescence immuno-assay, electroluminescence, optical fibers, and small particles.

10.1 Upconversion: Processes and Materials

10.1.1 Upconversion Processes

The principle of upconversion is schematically given in Fig. 10.1. This figure shows the energy level structure of an ion with ground state A and excited levels B and C. The energy differences between levels C and B and levels B and A are equal. Excitation

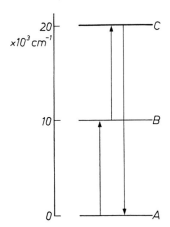

Fig. 10.1. The principle of upconversion. The infrared excitation radiation $(10\,000 \ \text{cm}^{-1})$ is converted into green emission $(20\,000 \ \text{cm}^{-1})$

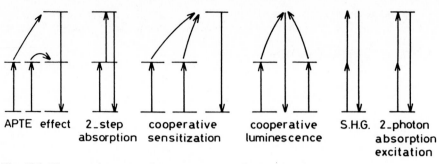

Fig. 10.2. The several upconversion processes according to Auzel [1]. See also text

occurs with radiation the energy of which corresponds to this energy difference, so that the ion is excited from A to B. If the life time of level B is not too short, the excitation radiation will excite this ion further from B to C. Finally emission from C to A may occur. Let us assume that the energy difference B-A and C-B is $10\,000$ cm^{-1} (corresponding to infrared excitation), then the emission is at $20\,000$ cm^{-1}, i.e. in the green. This is really anti-Stokes emission! It will also be clear that in this way we can detect infrared radiation visually. After this oversimplified introduction we now turn to a more serious consideration. Here we follow a treatment given by Auzel [1].

Actually there are many upconversion processes possible with widely different conversion efficiencies. Their energy schemes are given in Fig. 10.2. From left to right the following processes are shown:

– upconversion by energy transfer (Chapter 5), sometimes called the APTE effect (Addition de Photons par Transfers d'Energie). Here ions A transfer subsequently their excitation energy to another ion B which can now emit from a higher level.
– upconversion by two-step absorption (the example of Fig. 10.1) which needs only the B ion.
– upconversion by cooperative sensitisation: two ions A transfer simultaneously their excitation energy to ion C which has no energy level at the position of the excited level of A. Emission occurs from the excited level of C.
– cooperative luminescence: two A ions combine their excitation energy to one quantum which is emitted (note, however, that there is no real emitting level).
– second harmonic generation (frequency doubling) in which the frequency of the irradiated light is doubled (without any absorption transition taking place).
– two-photon absorption in which two photons are simultaneously absorbed without using any real intermediary energy level at all. The emission occurs by one photon from the excited energy level.

Table 10.1 gives some idea of the efficiency of the upconversion processes [2]; the efficiencies given relate to normalized incident power (1 W cm^{-2}). Also some examples of systems which show the relevant process are given. Table 10.1 together with Fig. 10.2 show that the higher-efficient processes require energy levels which are resonant with the incoming or outgoing radiation. This is not the case for the latter three, which will not be considered further.

Table 10.1 The different two-photon upconversion processes, their mechanism, their efficiency normalized to incident power (1 W cm^{-2}), and an example, after Refs [1] and [2]. Compare Fig. 10.2

Mechanism	Efficiency	Example
Sequential energy transfer (APTE)	10^{-3}	$YF_3 : Yb^{3+}, Er^{3+}$
Two-step absorption	10^{-5}	$SrF_2 : Er^{3+}$
Cooperative sensitization	10^{-6}	$YF_3 : Yb^{3+}, Tb^{3+}$
Cooperative luminescence	10^{-8}	$YbPO_4$
Second harmonic generation	10^{-11}	KH_2PO_4
Two-photon excitation	10^{-13}	$CaF_2 : Eu^{2+}$

10.1.2 Upconversion Materials

a. Materials with Yb^{3+} and Er^{3+}

The first example of upconversion was reported in 1966 by Auzel for the couple Yb^{3+}, Er^{3+} in $CaWO_4$ [3]. Figure 10.3 shows the energy level schemes involved. Near-infrared radiation (970 nm) is absorbed by Yb^{3+} ($^2F_{7/2} \rightarrow \, ^2F_{5/2}$) and transferred to Er^{3+}, so that the $^4I_{11/2}$ level of Er^{3+} is populated. During the lifetime of the $^4I_{11/2}$ level a second photon is absorbed by Yb^{3+} and the energy transferred to Er^{3+}. The Er^{3+} ion is now raised from the $^4I_{11/2}$ to the $^4F_{7/2}$ level. From here it decays rapidly and nonradiatively to the $^4S_{3/2}$ level from which a green emission occurs ($^4S_{3/2} \rightarrow \, ^4I_{15/2}$). In this way green emission is obtained from near-infrared excitation.

Since two infrared quanta are required to produce one green quantum, the emission intensity will increase quadratically with the density of the infrared excitation. This has been observed, and is proof of the two-photon character of the excitation. Efficiencies are therefore only useful if the excitation density is given.

In Table 10.2 the efficiency of the green emission intensity of Yb^{3+}, Er^{3+}-codoped host lattices under infrared excitation is given [4]. The excitation density is the same, as are the activator concentrations. It is seen that the efficiency depends strongly on the choice of the host lattice. That of α-$NaYF_4$ yields very efficient upconversion materials [5]. Oxides are less suitable than fluorides, since lifetimes in oxides are shorter than in fluorides due to a stronger interaction between the luminescent ion and its surroundings (Sect. 2.2). If the lifetime of the intermediary $^4I_{11/2}$ level is decreased, the total efficiency of the upconversion process will also decrease.

These materials are therefore able to convert infrared into green light. As an example, a GaAs diode (see Sect. 10.4) which yields infrared emission can be covered with a Yb^{3+}, Er^{3+}-codoped fluoride layer and yield green emission. However, there is no real advantage in this system, since a GaP diode yields green light directly. Although the efficiency of the GaAs diode is higher, the efficiency of the upconversion process is so low that the combination cannot really compete with the single GaP diode.

b. Materials with Yb^{3+} and Tm^{3+}

Figure 10.4 shows the relevant energy levels. Infrared radiation can be converted into

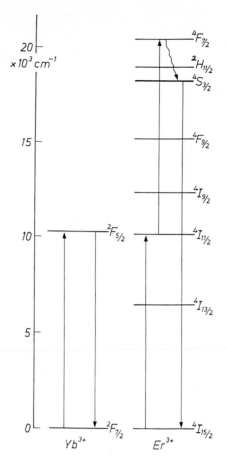

Fig. 10.3. Upconversion in the Yb^{3+}, Er^{3+} couple. Exciton is into Yb^{3+}, emission from the $^4S_{3/2}$ level of Er^{3+}

blue emission by a three-photon upconversion process. After one transfer step the $^3H_5$ level of Tm^{3+} is populated. This decays rapidly to the $^3F_4$ level. The second transfer step raises the Tm^{3+} ion from $^3F_4$ to $^3F_2$ which decays to $^3H_4$. Subsequently the third transfer step raises the Tm^{3+} ion from $^3H_4$ to $^1G_4$ which yields a blue emission. Its intensity increases linearly with the third power of the excitation density.

For Er^{3+} Auzel has described upconversion processes using summations of up to five photons [1]. In this way 970 nm radiation is converted into 410 nm radiation. Several of the transfer steps are not resonant, so that energy is lost to phonons in the transfer process. The first transfer step in the couple Yb^{3+}, Tm^{3+} is an illustrative example (see Fig. 10.4).

c. Materials with either Er^{3+} or Tm^{3+}
From Sect. 10.1.1 it is clear that materials doped with only one ion can only show

Table 10.2 Normalized green emission intensity under infrared excitation of Yb^{3+}, Er^{3+}-codoped host lattices [4]

Host lattice	Intensity
α-$NaYF_4$	100
YF_3	60
$BaYF_5$	50
$NaLaF_4$	40
LaF_3	30
La_2MoO_8	15
$LaNbO_4$	10
$NaGdO_2$	5
La_2O_3	5
$NaYW_2O_6$	5

upconversion with reasonable efficiency by the two- or more-step photon absorption process (Fig. 10.1). Two examples are given in Fig. 10.5. On the left-hand side the Er^{3+} ion converts 800 nm radiation into 540 nm radiation, on the right-hand side the Tm^{3+} ion converts 650 nm radiation into 450 and 470 nm radiation.

For application (see below) the ions are irradiated with a laser diode (Sect. 10.4): Er^{3+} absorbs 800 nm from an AlGaAs laser diode, Tm^{3+} absorbs 650 nm from newly developped laser diodes. Figure 10.5 shows that the Er^{3+} ion reaches the green-emitting $^4S_{3/2}$ level by subsequently absorbing two photons. Two different pathways are indicated; their relative importance depends on the ratio of the different transition rates. Also the Tm^{3+} ion reaches the blue emitting $^1D_2$ and $^1G_4$ levels by a two-step process. In both cases a quadratic dependence of emission intensity upon excitation power has been observed [6].

In fluoride glasses, these ions show a high upconversion efficiency, in contrast to silicate glasses (see also Sect. 10.1.2.a). A suitable glass is ZBLAN ($53ZrF_4$, $20BaF_2$, $4LaF_3$, $3AlF_3$, $20NaF$). Figure 10.6 gives the upconversion emission of ZBLAN glass with Er^{3+} under 800 nm irradiation. The $^2G_{9/2} \rightarrow {}^4I_{15/2}$ emission is very weak because of the fast nonradiative decay from $^2G_{9/2}$ to lower levels (see Fig. 10.5). The weak $^4F_{9/2} \rightarrow {}^4I_{15/2}$ emission is due to a small amount of nonradiative decay from $^4S_{3/2}$ to $^4F_{9/2}$.

The potential application of this glass is in high-density optical recording (used, for example, in compact disc players). In such devices the information density increases with decreasing size of the focus spot of the laser. This size varies inversely quadratically with the wavelength. Since the available diode lasers emit in the near infrared, there is a considerable amount of research going on to obtain a blue-emitting diode laser. There are three possibilities, viz.

(1) the search for a blue-emitting laser diode which does not exist at the moment, but may be developed on the basis of zinc sulfide. The realization of such a laser seems to be near.

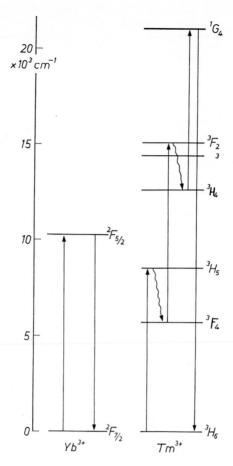

Fig. 10.4. Upconversion in the Yb^{3+}, Tm^{3+} couple. The emitting $^1G_4$ level of Tm^{3+} is reached after three successive energy transfer steps from Yb^{3+}

(2) frequency-doubling of the near-infrared laser radiation by second harmonic generation (see Fig. 10.2); materials for this purpose are $KNbO_3$ and $K_3Li_{1.97}Nb_{5.03}O_{15.06}$ [7].

(3) based upon the above-mentioned two-step absorption processes a fluoride-glass fiber can be made which acts as an upconversion laser, pumped by a laser diode.

d. Concluding Remarks

The materials mentioned above show that especially the trivalent rare earth ions are very suitable for upconversion processes. This is not surprising in view of their energy level schemes (see Fig. 2.14). These show often a wealth of intermediary levels. Upconversion has, however, also been observed for other ions. Examples are $5f^n$ ions (U^{4+} and Np^{4+} in $ThBr_4$ [8]), and transition metal ions ($MgF_2 : Ni^{2+}$).

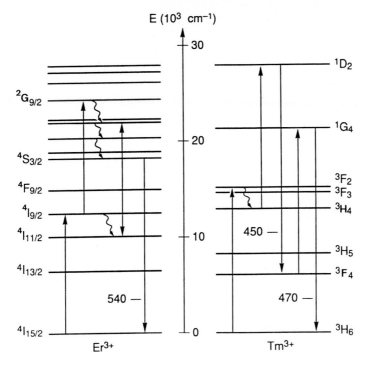

Fig. 10.5. Upconversion in one ion, illustrated on Er^{3+} and Tm^{3+}

The positive role of upconversion in the field of materials has already been outlined above. There are also negative aspects of upconversion, viz. saturation effects. This is due to the fact that upconversion implies a transition from an excited state upwards. If we are interested in the emission from this specific excited state, either in view of its luminescence or its stimulated emission, we have to consider the fact that upconversion will decrease the population of this level, so that the intensity of the emission of interest decreases. This will especially be the case for higher activator concentrations and/or high excitation densities (saturation effect).

Part of the saturation effects in projection-television phosphors (see Sect. 7.3.4) can be ascribed to an upconversion process of the type shown in Fig. 10.7. This is very similar to the Auger processes mentioned for semiconductors (see Sect. 4.6). Often this type of upconversion prevents a material from becoming a good laser material. If the stimulated emission radiation is reabsorbed by ions which are still in the excited state, the laser efficiency drops. From Fig. 10.7, it becomes clear that the upconversion process, which is, in this case, usually called excited state absorption, influences the population inversion in a negative way: considering the two ions in Fig. 10.7, the population inversion is complete before the upconversion occurs, but after upconversion and nonradiative decay to the emitting state, the population

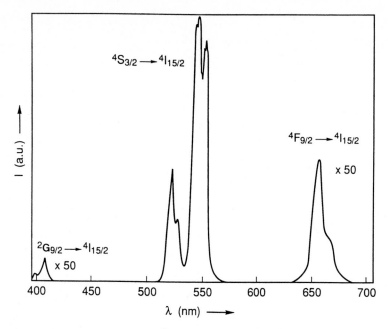

Fig. 10.6. The emission spectrum of Er^{3+}-doped ZBLAN glass in the visible spectral region under infrared excitation. Reproduced with permission from Ref. [6]

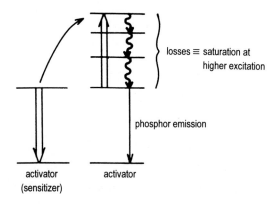

Fig. 10.7. Upconversion as a quenching process of the luminescence in a phosphor

inversion has decreased to 50%. For this reason excited-state absorption studies are popular among those who investigate potential laser materials.

For recent developments the reader is referred to the Proceedings of the International Conference in Luminescence 1993 (J. Luminescence, (1994) 60/61).

10.2 The Luminescent Ion as a Probe

The use of luminescent ions as a probe does not belong to the field of industrial application of luminescent materials, but should be considered as an application in the field of research and characterisation of materials. The basic idea is that the luminescence properties of an ion tell us something about the ion itself and also about its surroundings in the host lattice. The dangerous side of this use of luminescence, a side which is often overlooked, is that only luminescent ions can be monitored. However, it may well be that the material contains the specific ion but that it does not or only partly luminesce under the given circumstances. It is therefore important to know whether the ion has really been excited or not, and whether all ions show emission or not.

The luminescence of ions can be used as a tool for their chemical analysis. Fassel et al. [9] have described the techniques of optical atomic emission and X-ray excited optical luminescence for chemical analysis of rare earth elements. It is essential that the elements that are to be analysed are excited with high efficiency, and luminesce with high quantum efficiency. Under these circumstances analyses can be made in the ppm region, and sometimes even lower. An illustrative example is $GdAlO_3$ which shows always Cr^{3+} emission upon Gd^{3+} excitation [10]). The excitation energy on the Gd^{3+} ion migrates over the Gd^{3+} sublattice (Sect. 5.3.1) and is captured efficiently by the Cr^{3+} impurities which were present in the starting material containing aluminium. In the same way the rutile modification of TiO_2 shows always Cr^{3+} emission upon excitation into the band-band transition of rutile. The created free charge carriers (Sect. 3.3.9) recombine on the Cr^{3+} impurities and make it very hard to investigate the intrinsic rutile luminescence [11]. These experiments show that aluminium- and titanium compounds often contain some chromium. In addition the excitation routes in $GdAlO_3$ and TiO_2 are very efficient.

How luminescent ions can function as a probe of their surroundings, is easily understood from a consideration of Fig. 3.10. The Eu^{3+} ion shows very different emissions in $NaGdO_2$ and $NaLuO_2$. If the crystal structures were not known, one could find from these spectra that the Na^+ and Ln^{3+} ions (Ln = Gd, Lu) are ordered, and that the Eu^{3+} coordination in $NaLuO_2$ has inversion symmetry and that in $NaGdO_2$ not.

In this connection, it is important to note that luminescence yields structural information of a completely different character from that obtained by diffraction (X-rays, neutrons). The latter detects long-range order, i.e. the total crystal structure. The former yields only information on the surroundings of the luminescent ion, and can therefore probe short-range order.

As an example we mention the scheelites Y_2SiWO_8 and Y_2GeWO_8 [12]. X-ray diffraction does not reveal superstructure reflections due to a crystallographic order on the tetrahedral sites. The luminescence of a small amount of Eu^{3+} in these compounds yields other information. The silicon-containing compound shows sharp emission lines indicating a considerable amount of short-range order; however, the germanium containing compound shows emission lines which are an order of magnitude broader indicating a much lower degree of order (Fig. 10.8).

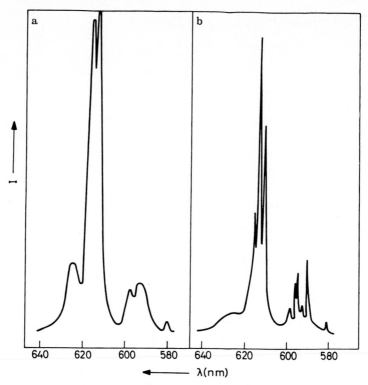

Fig. 10.8. The emission spectra of the Eu^{3+} ion in Y_2GeWO_8 (a) and Y_2SiWO_8 (b)

Often luminescent ions are used to characterise the structure of glasses. In our opinion such measurements yield only information on the surroundings of the luminescent ion in the glass, and not on the glass structure. In addition it should be kept in mind that the luminescent ion, often a network modifier, may locally perturb the glass structure strongly.

Boulon et al. [13,14] used the luminescence of Cr^{3+} in order to investigate the crystallization of a glass and to characterize glass ceramics. Use is made of the fact that the Cr^{3+} ion prefers the crystalline above the amorphous phase. Figure 10.9 shows one of their results. The glass has composition 52% SiO_2, 34.7% Al_2O_3, 12.5% MgO and 0.8% Cr_2O_3. Curve (a) gives the emission spectrum of the glass. It is due to Cr^{3+} ions. Some of these give ^2E emission (around 692 nm), but most of them $^4T_2$ emission (the band around 850 nm) (see also Sect. 3.3.4a). Curve (c) gives the emission of this glass after 10 min heating at 950°C. This is the temperature where crystallization occurs. Now the ^2E emission dominates and has also become sharper. Analysis shows that this emission is mainly due to Cr^{3+} in $MgAl_{2-x}Cr_xO_4$ crystals with 40 nm diameter. In the crystal the inhomogeneous broadening (Sect. 2.2) is less, so that the ^2E emission sharpens relative to that in the glass. In the spinel structure the crystal field on Cr^{3+} is so large that only ^2E emission occurs. The broad band in

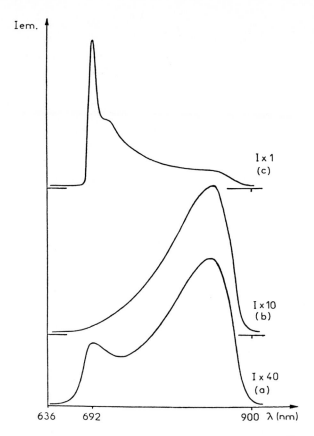

Fig. 10.9. Luminescence spectroscopy of glass crystallization. (a) the Cr^{3+} emission of the glass (^2E and $^4T_2$ emission); (b) the emission of the glass containing $MgCr_2O_4$ crystallites; (c) the emission of the glass containing $MgAl_{2-x}Cr_xO_4$ crystals. Note the large increase in luminescence intensity. See also text. Reproduced with permission from Ref. [13]

curve (c) is due to Cr^{3+} ions which are still in the glass where the crystal field is lower.

Curve (b) is the emission spectrum of a glass annealed at somewhat lower temperature. It has been shown that this glass contains very small $MgCr_2O_4$ crystallites, so that the conclusion follows that the Cr^{3+} ions in the glass cluster together and form $MgCr_2O_4$ crystallites as the very first crystallization step. Later the Cr^{3+} ions are redistributed to form $MgAl_{2-x}Cr_xO_4$, because particle growth occurs by assimilation of Al^{3+} ions.

Phase transitions in crystalline materials have also been followed by luminescence measurements. Examples are the transition crystalline → liquid crystalline in octa-n-dodecoxy substituted phthalocyanine [15] where the luminescence disappears, and the para- to ferroelectric transition in Cr^{3+}-doped $Li_2Ge_7O_{15}$ [16], where the ^2E emission lines are split by the phase transition.

Examples of probing the surroundings are the following:

- in $CaF_2 : Er^{3+}$ the luminescent ion has been found to occupy up to 20 distinct lattice sites [17]. This is due to the fact that the Er^{3+} ion has an excess positive charge relative to Ca^{2+} and requires charge compensation. In the fluorite structure this can occur in many ways, so that many centers appear. This has been found for many trivalent ions in fluorite. Examples of these centers are Er^{3+} without nearby charge compensator, Er^{3+} associated with O^{2-}, with interstitial F^-, associates of Er^{3+} and several F^- ions. By the application of site-selective spectroscopy it is possible to study the thermodynamics of the defect-defect interactions.

- in $CaSO_4$ the Eu^{3+} and other trivalent luminescent ions form associates with V^{5+} : $(Eu_{Ca}^{\bullet}.V_S')^x$. This follows directly from the emission spectra. Excitation into the vanadate group results in Eu^{3+} emission, because the V-Eu distance is short (nearest neighbors, see Sect. 5.3.2) [18]. Such samples can be considered as solutions of "molecules" $EuVO_4$ into $CaSO_4$. Unfortunately the solubility is low. Otherwise a cheap red-emitting lamp phosphor would have been found (see Sect. 6.5).

- in $CsCdBr_3$ the Cd^{2+} ions form linear chains. If three Cd^{2+} ions are replaced by two Ln^{3+} ions, they form a linear cluster $(Ln_{Cd}^{\bullet} \cdot V_{Cd}'' \cdot Ln_{Cd}^{\bullet})^x$. If $Ln = Tb$, such a cluster yields green emission from the $^5D_4$ level because of cross relaxation (Sect. 5.3.1), so that even for low Tb^{3+} concentrations the blue emission intensity is low [19].

- the way in which trivalent lanthanide ions are coordinated on a silica surface can be deduced from their emission spectra [20]. The Ln^{3+} ion is bound directly via Si–O-bonds to the silica and is, on the other-side, coordinated by 4 water molecules.

- the way in which the vanadate group is attached to the silica surface in the industrially important catalyst SiO_2–V can be found from luminescence spectroscopy [21]. These materials show a peculiar luminescence: the vanadate emission band shows vibrational structure at low temperature with a spacing between the individual lines of nearly 1000 cm^{-1} (see Fig. 10.10). This indicates the presence of a very short V-O bond. The decay time is very long. All this shows that the V^{5+} ion is bonded to the silica surface with three Si–O–V bonds, whereas one very short V–O bond sticks out of the surface (V–O distance ~ 1.56 Å). It is this V–O bond which is involved in catalytic processes (Fig. 10.11).

10.3 Luminescence Immuno-Assay

10.3.1 Principle

The luminescence of rare earth complexes can be applied in immunology, which is a method for determining biological species, especially for clinical applications. The method is superior to many other methods as far as sensitivity and specificity

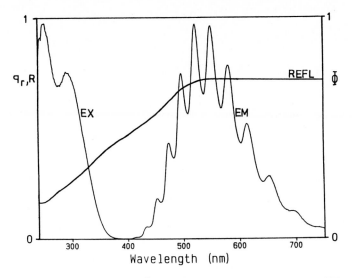

Fig. 10.10. The spectroscopy of SiO_2-V^{5+}. REFL: diffuse reflection spectrum; EX: excitation spectrum of the luminescence; EM: emission spectrum at low temperatures showing vibrational structure. After M.F. Hazenkamp, thesis, University Utrecht, 1992

Fig. 10.11. The vanadate group on the silica surface

are concerned. Although this book deals with solid materials only, this application is mentioned here in view of the interesting aspects of the luminescent species involved. Their properties have also a strong analogy to those of luminescent ions in solids. The whole topic has recently been reviewed by Sabbatini et al. [22].

The immunological method based on the use of luminescent labels is usually called fluoroimmunoassay [23]. Here we use the term luminescence immunoassay for the same reasons given in Sect. 6.2 for luminescent lighting. The luminescent label is coupled chemically to an antibody which binds in a specific way to a given biomolecule or organism. In this way the presence of luminescence can be related to the presence of certain molecules or organisms.

Usually the samples have their own luminescence. Therefore rare-earth labels are used. The background luminescence of biological materials is usually short-lived, whereas ions like Eu^{3+} and Tb^{3+} have emitting states with long lifetimes (Sect. 3.3.2.). Therefore these two types of emission can be easily separated.

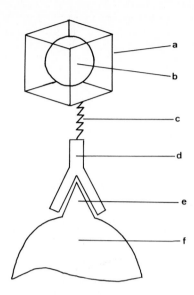

Fig. 10.12. Schematic representation of luminescence immunoassay with a rare-earth cryptate. a: cryptand; b: luminescent rare earth ion; c: connection to antibody; d: antibody; e: antigene; f: biomolecule

Since the whole determination is carried out in aqueous media, the rare-earth ion has to be shielded from its direct environment, because otherwise its luminescence is strongly quenched by the water molecules (Sect. 4.2.1). There are several ways of preventing this, as will be discussed below. In all of these a cage is built around the luminescent ion. Figure 10.12 shows the principle of luminescence immunoassay in a schematic way.

10.3.2 Materials

The commercial kits presently available make use of Eu^{3+} chelates. It is a complicated procedure, since two different chelates are used. More promising are complexes consisting of a rare earth ion and an encapsulating ligand. Several types of these ligands have been proposed, see Ref. [22]. Here we restrict ourselves to the cryptands. A cryptand is a ligand which forms a cage around a metal ion; the whole complex is called a cryptate. An example of such a cryptand is given in Fig. 10.13. The $(Eu \subset 2.2.1^{3+}$ and $(Tb \subset 2.2.1)^{3+}$ cryptates ($\subset$ stands for encapsulation) show luminescence in aqueous solution, but their quantum efficiencies are low, especially in the case of the Eu^{3+} ion (excitation in the charge-transfer and $4f$-$5d$ transition, respectively). The nonradiative losses are due to multiphonon emission involving the water molecules and, in the case of Eu^{3+}, the low energy position of the charge-transfer state (Sect. 4.2.2).

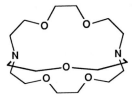

Fig. 10.13. 2.2.1 cryptand

Nevertheless the cryptand shields the rare earth ion from the aqueous medium, since the free rare earth ions have still lower efficiencies in aqueous solution. In the latter case 9 to 10 water molecules coordinate the rare earth ion, whereas in the 2.2.1 cryptate only 3 water molecules are still able to interact with the rare earth ion.

Considerable improvements have been obtained by using cryptands containing organic groups with strong ultraviolet absorption. The prototype is the bpy.bpy.bpy cryptand (bpy = 2,2'-bipyridine). Fig. 1.9 shows the relevant cryptate. This cryptate shows a high output of the lanthanide emission, because ultraviolet radiation is strongly absorbed in the bpy groups. The absorbed energy is subsequently transferred to the rare earth ion, from where emission occurs. There is a strong analogy with $YVO_4 : Eu^{3+}$ (Sect. 5.3.2). The VO_4^{3-} group as well as the bpy complex are efficient sensitizers (Sect. 5.1) of the rare earth emission. In the language of photochemists this is called the antenna effect [22].

However, nonradiative losses also occur here, because the cryptand does not shield the rare earth ion completely. Even for Tb^{3+} the quantum efficiency is low. This is due to the presence of a charge-transfer state between Tb^{3+} and the cryptand which leads to nonradiative return to the ground state (Sect. 4.5). Note again the analogy with $YVO_4 : Tb^{3+}$, which also does not luminesce efficiently due to quenching via a charge (or electron) transfer state (Sect. 4.5).

These findings also suggest the way to improvement: Tb^{3+} is to be preferred over Eu^{3+} because its larger gap between the emitting level and the next lower level (see Fig. 2.14) makes its emission less sensitive to the presence of water molecules (compare also Sect. 4.2.1). However, the use of Tb^{3+} will only lead to efficient luminescence if higher excited configurations (like charge-transfer states) are at high energy. This has been realized by using another type of macrocyclic ligand which shields better from the water molecules and does not introduce harmful excited states. This ligand is bpy-branched triazacyclononane [22]. Up till now the maximum quantum efficiencies obtained for these types of complexes are 20% for Eu^{3+} and 40% for Tb^{3+}.

In closing we note that a very different way to realize luminescence immunoassay is the use of commercial phosphors. By using $YVO_4 : Eu^{3+}$, for example, it has been possible to achieve acceptable results. The powder particles are bound to the antibody and are in this way connected to the object of study. The quantum efficiencies are high, and shielding from water molecules is no longer a problem because the luminescent species are in the solid. However, even the finest particles are very large relative to the molecular scale on which the cryptate operates.

Some research is known in which the potential of the luminescence of "caged" rare earth ions was investigated in a very general way. Also zeolites have been included [24]. Zeolites are solid materials which contain internal holes which can contain ions or molecules. They are widely applied in catalysis and their cost is low. Up till now the results on these systems with "caged" ions or molecules are not too promising. Nevertheless it is clear that the use of organic groups as absorbing features has a large advantage over inorganic groups, since their absorption strength is easily one order of magnitude higher.

10.4 Electroluminescence

10.4.1 Introduction

When a luminescent material can be excited by application of an electric voltage, we speak of electroluminescence. In order to convert electric energy from the applied voltage into radiation, three steps have to be considered: excitation by the applied field, energy transport to the luminescent center, and emission from this center. Acoording to the voltage applied, one can distinguish between low-field or high-field electroluminescence. Light-emitting diodes, where energy is injected into a p-n junction, are typical of low-field electroluminescence. The applied voltage is characteristically a few volts. High-field electroluminescence requires electric fields of 10^6 Vcm^{-1}. Materials based on ZnS are popular in this type of electroluminescence. Low-field electroluminescence operates usually with direct current, whereas high-field electroluminescence operates usually with alternating current (ACEL). In this paragraph we will first consider low-field electroluminescence (with applications like light-emitting diodes and laser diodes, also called semiconductor lasers), and subsequently high-field electroluminescence which shows a potential for display (thin-film electroluminescence). For more elaborate treatments the reader is referred to several chapters in a recent book edited by Kitai [25].

10.4.2 Light-Emitting Diodes and Semiconductor Lasers

Semiconductors can be classified as n-type or p-type depending on the nature of the dopant. For our purpose we consider a p-n junction, i.e. the interface between a piece of n-type and a piece of p-type semiconductor. The band structure and the electron distribution around the junction are drawn schematically in Fig. 10.14.

A voltage is applied in such a way that electrons are supplied to the n-type side of the junction (forward bias). As a consequence electrons in the conduction band of the n-type semiconductor fall into the holes of the valence band of the p-type semiconductor (see Fig. 10.15). In certain junctions, depending on the nature

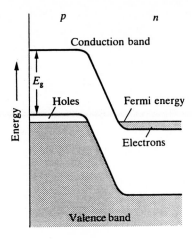

Fig. 10.14. Energy level structure of a p-n junction

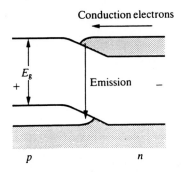

Fig. 10.15. As Fig. 10.14 with applied forward bias. Figs 10.14 and 10.15 are reproduced with permission from Ref. [38]

of the semiconductor, the energy which is involved in this transitions is emitted as radiation. This is especially the case for direct semiconductors which are defined as semiconductors with an optically allowed band-to-band transition. In this way we have obtained a light-emitting diode. These are applied in electronic displays.

Such a light-emitting diode is not yet a laser. However, it is easy to use the radiative electron-hole recombination in a p-n junction as the basis of a laser. Population inversion can be realized by rapidly taking away the electrons which have fallen into the holes of the valence band. These lasers are nowadays widely used (optical communication, compact disc player). An additional large advantage is their small dimension (< 1 mm).

Using GaAs it is easy to produce infrared emission from a p-n junction. The upconversion of this radiation into visible radiation was discussed in Sect. 10.2. By incorporating phosphor in GaAs, the band gap increases and the emission shifts to

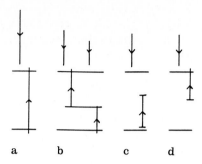

Fig. 10.16. Impact processes in a solid. a: Band-to-band ionization. b: Two-step band-to-band ionization. c: Impact excitation. d: Impact ionization. After J.W. Allen [27]

the visible. For example, $GaAs_{0.6}P_{0.4}$ shows red-light emission, and GaP green. The former is still a direct semiconductor, the latter however is an indirect semiconductor. Since the optical transition is now forbidden, i.e. slow, the radiationless processes become of more importance, so that this diode has a low efficiency.

The semiconductors ZnS and ZnSe have still higher band gaps which decreases the radiative rate further. Also it has been hard to produce n- as well as p-type material. Although some progress has been made, the success of diodes based on II-VI semiconductors is still restricted [26].

10.4.3 High-Field Electroluminescence

The discovery of ACEL in powders by Destriau goes back to 1936. Nevertheless our understanding of the fundamental processes is restricted as reviewed by Allen [27]. This author stresses the analogy between high-field electroluminescence and the phenomena in a gas discharge lamp. In such a lamp, atoms are excited or ionized by impact with energetic electrons which gain energy from an applied voltage (compare also Fig. 6.1 and Sect. 6.2). That the efficiency of such a system can be high, follows from the success of luminescent lighting. The solid-state analogue (high-field electroluminescence) did, however, not bring such high efficiencies.

Electrons (or holes) in a solid can be accelerated by an electric field. However, they can easily lose energy by phonon emission (i.e. by exciting lattice vibrations). Therefore a high field is necessary, so that the gain from the field exceeds the loss to the phonons. Since the path-length in a solid is small, the luminescent center concentration should be high; a limit may be set by concentration quenching. The impact processes to be considered are schematically depicted in Fig. 10.16.

(a) band-to-band impact ionization creates electrons and holes which can recombine radiatively through a luminescent center. The disadvantage is that the current increases very rapidly with voltage (this prevents stable operation), and that the fields required are very high.

(b) in the two-step band-to-band impact ionization, an incident hot carrier ionizes a deep level and then another hot carrier raises an electron from the valence band to the deep level. In this way free electrons and holes are produced. In ZnS and ZnSe, appreciable carrier generation can occur at fields an order of magnitude lower than required for single-step band-to-band ionization. It is attractive to have a high concentration of the impact-ionizable centre and a lower concentration of luminescent centers in order to prevent concentration quenching.

(c) impact excitation of the luminescent center. This seems to be the mechanism for ZnS : Mn^{2+} which is used in commercially available devices. The power efficiency η can be estimated as

$$\eta \sim h\nu\sigma N/eF \tag{10.1}$$

Here F is the electric field, $h\nu$ the emitted energy, and $(sN)^{-1}$ the distance which a hot electron covers between two impacts (σ is the cross section and N the optimum luminescent center concentration). With the typical values $h\nu = 2$ eV, $s = 10^{-16}$ cm^2, $N = 10^{20}$ cm^{-3}, and $F = 10^6$ Vcm^{-1}, it follows that $\eta \sim 2\%$, in agreement with experiment.

(d) Impact ionization of a luminescent center seems to occur in the electroluminescence of rare-earth doped SrS and CaS.

At the moment it is difficult to see how a considerable improvement in high-field electroluminescence efficiency can be obtained.

In thin-film electroluminescence, a transparent front electrode and an opaque back electrode are used. The former may be a thin layer of indium-tin oxide, the latter a thin layer of aluminium. In the early 1970s a MISIM device (metal-insulator-semiconductor-insulator-metal) with a ZnS : Mn^{2+} electroluminescent layer was shown to maintain bright luminescence for thousands of hours. The insulators can be selected from a large groups of oxides.

The preparation method of the thin layers and the electroluminescent materials available are reviewed in Ref. [28]. Among the techniques used are sputtering, vacuum evaporation, metal-organic chemical-vapor deposition (MOCVD) and atomic layer epitaxy (ALE).

The most successful material is ZnS : Mn^{2+}. Its emission has a yellow colour (see also Fig. 10.17). The transition involved is the well-known $^4T_1 - {}^6A_1$ transition (Sect. 3.3.4c). The optimum Mn concentration is about 1 mole %.

In order to realize multicolor electroluminescent devices there has been world-wide research into other materials. None has been found to be as efficient as ZnS : Mn^{2+}. As examples, we can mention ZnS : LnF^{3+} (Ln = rare earth) films where, for example, Ln = Tb^{3+} gives green and Ln = Sm^{3+} red electroluminescence, and MS(M = Ca,Sr) : Ce^{3+} or Eu^{2+} (green and red emission, respectively). Many other proposals can be found in the literature.

Thin-film electroluminescent (TFEL) devices can be used for display purposes [29]. Since 1983, Helsinki airport uses such a display based on ZnS:Mn^{2+} in the arrival hall. At the moment multi-color devices are being brought out on the market, and are expected to take about 5% of the flat-panel display market in the coming years (the greater part still being color liquid-crystal displays). The red and green

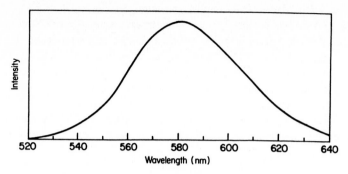

Fig. 10.17. The electroluminescence emission spectrum of ZnS : Mn^{2+}

are produced by the emission of ZnS : Mn^{2+} using filters (compare Fig. 10.17). The problem is in the blue, and a good material is still lacking. The SrS : Ce films seem to be the best solution to fill in the gap. More recently cerium-activated alkaline earth thiogallates were proposed [29].

10.5 Amplifiers and Lasers with Optical Fibers

The first lasers with optical fibers were proposed in 1964 and realized in 1974 [30]. For a few years this field of application has been exploding, since large successes were obtained in the production of useful amplifyers in the spectral area around 2 μm, a wavelength of recognised interest for long- and medium-distance optical communication. Silica fibers doped with Er^{3+} are commercially available for this purpose.

Because we are concentrating the discussion on the Er^{3+} ion, its energy level scheme, as far of importance, is given in Fig. 10.18. The Er^{3+} ion can be pumped with 0.98 or 1.48 μm radiation from a semiconductor laser (Sect. 10.4), i.e. in the $^4I_{15/2}$ – $^4I_{11/2}$ or $^4I_{13/2}$ transition, respectively. (Stimulated) emission will occur at 2.5 μm ($^4I_{11/2} \rightarrow {}^4I_{13/2}$) or 1.5 μm ($^4I_{13/2} \rightarrow {}^4I_{15/2}$).

In Sect. 3.6 it was argued that a three-level laser is not so easy to operate as a four-level laser. The reason for this is that a three-level laser requires saturation of the pumping transition, whereas a four-level laser requires only compensation of the optical losses. Fibers, however, have a large advantage over crystals or other bulky materials, viz. their geometric configuration (small diameter and large length). Such a configuration makes it very easy to satisfy laser conditions, even though this is not possible in a bulky sample [30,31].

The interest in Er^{3+}-doped fibers stems from the fact that silica as well as fluoride glasses show very low absorption in the emission region mentioned above (see Fig. 10.19). The silica glasses transmit the $^4I_{13/2} \rightarrow {}^4I_{15/2}$ emission best, the fluoride glasses the $^4I_{11/2} \rightarrow {}^4I_{13/2}$ emission. Figure 10.19 is an absorption spectrum; for

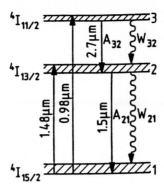

Fig. 10.18. Lower energy level scheme of Er^{3+}. Compare also Table 10.3

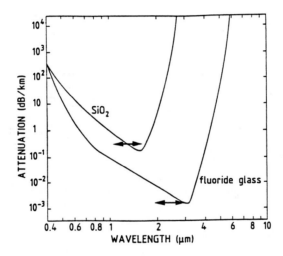

Fig. 10.19. The optical transmission spectra of undoped silica and fluoride glass

fibers the attenuation of the optical signal of a given wavelength is given in dB/km. An emission spectrum of the 2.7 μm $^4I_{11/2} \rightarrow {}^4I_{13/2}$ Er^{3+} emission of ZBLAN : Er^{3+} (compare Sect. 10.1.2c) is given in Fig. 10.20 for bulk and fiber. The considerable narrowing in the case of the fiber points to amplification by stimulated emission.

It is of interest to compare Er^{3+} in silica and in fluoride glass. Some data are given in Table 10.3 where the notation refers to the energy level scheme in Fig. 10.18. The radiative rates are about the same, but the nonradiative ones are very different. The latter is due to the difference between the maximum vibrational frequencies (1100 cm^{-1} for silica glass and 500 c^{m-1} for the fluoride glass) (compare also Sect. 4.2.1). The data show that Er^{3+} in silica will mainly give the $^4I_{13/2} \rightarrow {}^4I_{15/2}$ emission at 1.5 μm. Fluoride glasses will also give the 2.7 μm $^4I_{11/2} \rightarrow {}^4I_{13/2}$ emission.

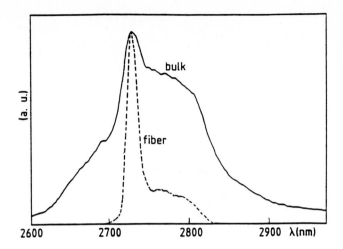

Fig. 10.20. Emission spectrum of a ZBLAN : Er^{3+} glass for a bulk sample and a fiber. After F. Auzel [30]

Table 10.3 Some data on the transitions of Er^{3+} in glasses. Compare also Fig. 10.18

Transition rate		Transition probability (s^{-1})	
Radiative	Nonradiative	Silica	Fluoride
A_{32}		10	10
	W_{32}	10^6	50
A_{21}		100	100
	W_{21}	250	2.10^{-4}

In conclusion, doped optical fibers bring large gains, even for optical transitions which cannot be used for that purpose in a bulk sample. They are now part of telecommunication nets, and most probably cannot be replaced.

10.6 Luminescence of Very Small Particles

The luminescence of small particles, especially of semiconductors, is a fascinating development in the field of physical chemistry, although it is too early to evaluate the potential of these particles for applications. The essential point is that the physical properties of small semiconductor particles are different from the bulk properties and from the molecular properties. It is generally observed that the optical absorption edge shifts to the blue if the semiconductor particle size decreases. This is ascribed to the quantum size effect. This is most easily understood from the electron-in-a-box model. Due to their spatial confinement the kinetic energy of the electrons increases. This results in a larger band gap.

As an example, we can mention colloids of ZnS. They can be prepared with variable particle size (down to 17-Å diameter, which corresponds to particles containing about 60 molecules of ZnS). The onset of the optical absorption shifts from 334 nm (large particles and bulk) to 288 nm (17-Å particles). The particles show luminescence. Also the emission maximum shifts to shorter wavelength if the particle size is decreased.

A recent, well-defined example of such clusters is the existence of CdS superclusters in zeolites [32]. The authors prepared very small CdS clusters in zeolites. In zeolite Y, for example, there are sodalite cages (5 Å) and supercages (13 Å). Well-defined clusters can be made using these cages. The zeolites were Cd^{2+}-ion exchanged and subsequently fired in H_2S. The resulting zeolite is white (note that CdS is yellow). The products were characterized by several means. It was shown that CdS is within the zeolite pore structure. There are discrete $(CdS)_4$ cubes in the sodalite cages. The cubes consist of interlocking tetrahedra of Cd and S. For high enough CdS concentrations these clusters are interconnected. As this interconnection proceeds, the absorption spectra shift in band edge from 290 to 360 nm. These materials show luminescence. Three different emissions have been observed, viz., a yellow-green one (ascribed to Cd atoms), a red one (ascribed to sulfur vacancies), and a blue one (ascribed to shallow donors). A very interesting aspect is that the vibrational mode responsible for the nonradiative transitions in these materials has a frequency of 500-600 cm^{-1}. This is higher than the highest phonon frequency in CdS. This indicates that interface and host (zeolite) phonons are responsible for these processes.

Recently the same group synthesized a compound $Cd_{32}S_{14}(SC_6H_5)_{36} \cdot DMF_4$ [33]. Its structure contains an 82 atom CdS core that is a roughly spherical piece of the cubic sphalerite lattice ~ 12 Å in diameter. The crystals of this compound emit a green luminescence with a maximum at about 520 nm. The corresponding excitation maximum is at 384 nm. Luminescence spectra at 6.5 K are given in Fig. 10.21.

A comparison of this excitation spectrum with the absorption spectrum of CdS (see Fig. 10.21) shows clearly that the small CdS cluster absorbs at much higher energy than the CdS bulk. Broad band emission from CdS is usually interpreted as luminescence due to defect centers (see Sect. 3.3.9a). However, the CdS core in $Cd_{32}S_{14}(SC_6H_5)_{36} \cdot DMF_4$ has no defects. Herron et al. [33] have ascribed the emission to a charge-transfer transition.

Since bulk CdS shows free-exciton emission at low temperatures [34], it is interesting to compare these results on CdS with the discussion in Sect. 3.3.9b on the transition from semiconductors to insulators. There it was shown that narrow-line free-exciton emission transforms into broad-band localized emission, if the amount of delocalization of the excited state decreases. Since the valence band to conduction band transition of CdS is in principle a $S^{2-} \rightarrow Cd^{2+}$ charge-transfer transition, this would bring the discussion on CdS in line with results from a different origin (see Sect. 3.3.9b). By all means the case of $Cd_{32}S_{14}(SC_6H_5)_{36} \cdot DMF_4$ is a nice example of luminescence research on a well-defined cluster showing the quantum-size effect.

An interesting development in this field is the report by Dameron et al. [35] of the biosynthesis of quantum-sized CdS crystals in the yeast cells Candida glabrata

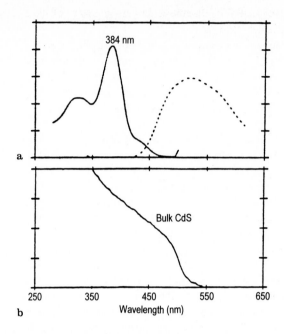

Fig. 10.21. A: Emission (broken line) and excitation spectra of the luminescence of $Cd_{32}S_{14}(SC_6H_5)_{36}.DMF_4$ at 6.5 K. B: Absorption spectrum of bulk CdS. After data in Ref. [33]

and schizosaccharomyces pombe. Exposed to Cd^{2+} ions these cells synthesize certain peptides with an enhanced sulfide production. Small CdS crystals are formed inside the cells. These crystallize in the rock salt structure (and not in the thermodynamically stable hexagonal configuration). The organism controls particle nucleation and growth, so that uniformly sized CdS particles of about 20 Å are formed. They show pronounced quantum-size effects. This is the first example of the biosynthesis of quantum-sized semiconductor crystallites. It constitutes a metabolic route for the detoxification of Cd^{2+}-infected living cells.

No doubt these semiconductor superclusters present a novel class of materials where the three-dimensional structure can be controlled. They present a challenge to synthetic and physical chemists. The most recent developments can be found in Refs [36] and [37].

References

1. Auzel F (1985) In Jezowska-Trzebiatowska B, Lengendziewicz J, Strek W (eds) Rare earth spectroscopy. World Scientific, Singapore, p 502
2. Auzel F (1973) Proc. IEEE 61: 758
3. Auzel F (1966) C.R. Ac. Sci (Paris) 262 :1016 and 263: 819
4. Sommerdijk JL, Bril A (1974) Philips Techn. Rev. 34: 24

5. Blasse G, de Pauw ADM (1970) unpublished
6. Oomen EWJL (1991) Adv. Mater. 3: 403
7. Ouwerkerk M (1991) Adv. Mater. 3: 399
8. Genet M, Huber S, Auzel F (1981) C.R. Ac. Sci. (Paris) 293: 267
9. DeKalb EL, Fassel VA (1979) In Gschneidner Jr KA, Eyring L (eds) Handbook on the physics and chemistry of rare earths Holland, Amsterdam, chap 37D; D'Silva AP, Fassel VA (1979) ibid. chap 37E
10. de Vries AJ, Smeets WJJ, Blasse G (1987) Mat. Chem. Phys. 18: 81
11. de Haart LGJ, Blasse G (1986) J. Solid State Chem. 61: 135
12. Blasse G (1968) J. Inorg. Nucl. Chem. 30: 2091
13. Boulon G (1987) Mat. Chem. Phys. 16: 301
14. Durville F, Champignon B, Duval E, Boulon G (1985) J. Phys. Chem. Solids 46: 701
15. Blasse G (1990) Adv. Inorg. Chem. 35: 319
16. Basun SA, Kaplyanskii AA, Feofilov SP (1992) Sov. Phys. Solid State 34: 1807
17. Tallant DR, Wright JC (1975) J. Chem. Phys. 63: 2074
18. Draai WT, Blasse G (1974) Phys. Stat. Sol.(a) 21: 569
19. Lammers MJJ, Blasse G (1986) Chem. Phys. Letters 126: 405; Berdowski PAM, Lammers MJJ, Blasse G (1985) J. Chem. Phys. 83: 476
20. Hazenkamp MF, Blasse G (1990) Chem. Mater. 2: 105
21. Hazenkamp MF, Blasse G (1992) J. Phys. Chem. 96 :3442; (1993) Research Chem. Intermediates 19: 343
22. Sabbatini N, Guardigli M, Lehn JM (1993) Coord. Chem. Revs 123 :201; Sabbatini N, Guardigli M (1993) Mat. Chem. Phys. 31: 13
23. Soini E, Hemmila I (1979) Clin. Chem. 25 :353
24. Bredol M, Kynast U, Ronda C (1991) Adv. Mater. 3: 361
25. Kitai AH (ed) (1993) Solid state luminescence. Chapman and Hall, London
26. Walker CT, DePuydt JM, Haase MA, Qiu J, Cheng H (1993) Physica B185 :27
27. Allen JW (1991) J Luminescence 48,49: 18; Chadha SS (1993) chap 6 in Ref. [25]
28. Leskelä M, Tammenmaa M (1987) Mat. Chem. Phys. 16: 349; Müller GO (1993) chap 5 in Ref. [25]
29. Morgan N (1993) Opto and Laser Europe 4 (March) p 26; (July) p 29; Mach R (1993) chap 7 in Ref. [25]
30. Auzel F (1993) In: Proceedings int. conf. defects insulating materials, Nordkirchen. World Scientific, Singapore, p 39; Snitzer E (1994) J. Luminescence, 60/61:145
31. Yen WM (1989) In Jezowska-Trzebiatowska B, Legendziewicz J, Strek W (eds) Excited states of transition metal elements. World Scientific Singapore, p 621
32. Herron N, Wang Y, Eddy MM, Stuckey GD, Cox DE, Moller K, Beitz T (1989) J. Am. Chem. Soc. 111 :530; Wang Y, Herron N (1988) J. Phys. Chem. 92: 4988
33. Herron N, Calabrese JC, Farneth WE, Wang Y (1993) Science 259: 1426
34. See, for example, Shionoya S (1966) In: Goldberg P (ed) Luminescence of inorganic solids. Academic New York, p 205
35. Dameron CT, Reese RN, Mehra RK, Kortan AR, Carroll PJ, Steigerwalt ML, Bras LE, Winge DR (1989) Nature (London) 338: 596
36. Bawendi MG, Carroll PJ, Wilson WL, Brus LE (1994), J. Chem. Phys. 96:946
37. Bhargava R, Gallagher D, Welker T (1994) J. Luminescence 60/61:275; Bhargava R, Gallagher D, Hong X, Nurmikko A (1994) Phys. Rev. Letters 72:416
38. Atkins PW (1990) Physical chemistry. Oxford University Press, Oxford, 4th ed

Appendix 1. The Luminescence Literature

The present text has clearly the character of an introduction into the field. Those who want to know more or to keep up with the luminescence literature do not have an easy task, since this literature is widely spread. The suggestions which follow originate from the authors experience and personal preference.

Books
– A.H. Kitai, editor, Solid State Luminescence; Theory, materials and devices, Chapman and Hall, London, 1993. A multi-authored book on many topics in the field of luminescence and luminescent materials. Many topics are discussed; the stress on electroluminescence is strong.

– B. Henderson and G.F. Imbusch, Optical Spectroscopy of Inorganic Solids, Oxford Science Publication, Clarendon Press, Oxford, 1989. An extensive treatment of luminescent centers in solids. Chapters on color centers, lasers and experimental techniques are included.

– S.W.S. McKeever, Thermoluminescence of Solids, Cambridge University Press, Cambridge, 1985. An excellent treatment of thermoluminescence and related phenomena.

– Proceedings of the Summer Schools on Spectroscopy in Erice (Sicily) edited by B. Di Bartolo and published by Plenum Press, New York. The first one dates from 1975; there has been a continuation every two years. Each book contains several chapters at an advanced level on all possible topics of spectroscopy and luminescence and (to a less extent) materials.

– Mat. Chem. Phys. 16 (1987) no. 3-4, a special issue devoted to new luminescent materials edited by G. Blasse. This issue contains several review papers with a stress on materials (X-ray phosphors, lamp phosphors, glasses, and electroluminescent thin films).

– K.H. Butler, Fluorescent Lamp Phosphors, The Pennsylvania State University Press, University Park, 1980. Interesting for its historical and application chapters. The book does not include the developments after the introduction of the rare earth phosphors.

– L. Ozawa, Cathodoluminescence, VCH, Weinheim, 1990.

– T. Hase, T. Kano, E. Nakazawa and H. Yamamoto, Phosphor Materials for Cathode-Ray Tubes, in Adv. Electr. Electron Physics (Ed. P.W. Hawkes) 79 (1990) 271. The latter two texts describe the theory and application of cathode-ray luminescence.

– Heavy Scintillators for Scientific and Industrial Applications (Eds F. De Notaris-tefani, P. Lecoq and M. Schneegans), Editions Frontières, Gif-sur-Yvette, 1993; proceedings of the workshop Crystal 2000. A good overview of the state-of-the-art in the scintillator field.

Journals

The Journal of Luminescence publishes fundamental papers on luminescence. Interesting are the issues which contain the proceedings of the International Conference on Luminescence (every three years) and of the Dynamics of the Excited State (DPC) meetings (every two years). A large number of relatively short papers show the state-of-the-art in the total field of luminescence throughout the years.

Many other fundamental papers in the field of luminescence can be found in the journals on solid state physics. Here we can mention Physical Review B (USA), Journal of Physics: Condensed Matter (UK), Journal of Physics and Chemistry of Solids, Journal of Experimental and Theoretical Physics (Russia), Physica Status Solidi (Germany).

The field of luminescent materials used to be covered by the Journal of the Electrochemical Society. This has changed completely. Such papers may nowadays be found in journals like the Journal of Solid State Chemistry, Materials Research Bulletin, Chemistry of Materials (USA), Journal of Materials Chemistry (UK), Journal of Materials Science, Japanese Journal of Applied Physics, and Journal of Alloys and Compounds. The latter journal publishes the proceedings of the international rare-earth conferences with many papers on the spectroscopy and luminescence of rare-earth containing materials.

The patent literature is left out of consideration.

Appendix 2. From Wavelength to Wavenumber and Some Other Conversions

Consider an optical center with two energy levels with energies E_1 and E_2 ($E_2 > E_1$). The center can be excited from the lower to the higher level by absorption of radiation. The frequency ν of this radiation is given by the well-known relation $\Delta E = E_2 - E_1 = h\nu$. The frequency is expressed in the unit s^{-1} (the number of vibrations per second). Spectroscopists more often use the wavenumber $\tilde{\nu} = \frac{\nu}{c}$ where c is the velocity of light. The unit of $\tilde{\nu}$ is cm^{-1}. Note that both ν and $\tilde{\nu}$ are linearly proportional to the energy.

Spectrometers often use the wavelength (λ) of the radiation. The wavelength follows from another well-known relation, viz. $\nu\lambda = c$. The wavelength is expressed in nm (10^{-9} m) or mm (10^{-6} m). Note that λ is not proportional to the energy.

Table A.2.1 gives for radiation of a given energy the color, wavelength, wavenumber and frequency. The visible area corresponds to $400 < \lambda < 700$ nm; ultraviolet radiation has $\lambda < 400$ nm, infrared radiation $\lambda > 700$ nm. In wavenumbers these regions are indicated as follows: $14.300 < \tilde{\nu} < 25.000$ cm^{-1} is visible, $\tilde{\nu} > 25.000$ cm^{-1} is ultraviolet, and $\tilde{\nu} < 14.300$ cm^{-1} is infrared.

Semiconductor physics often uses the electron volt (eV) as a measure of energy: 1 eV is the energy acquired by an electron when it is accelerated through a potential difference of 1 V. The energy of 1 eV corresponds to 8065.5 cm^{-1}.

Table A.2.1. Conversion of spectral units for radiation of certain colors.

Wavelength λ (nm)	Frequency $\nu(10^{14}s^{-1})$	Wavenumber $\tilde{\nu}(10^3 cm^{-1})$	Energy (eV)	Colour
1000	3.0	10.0	1.24	infrared
700	4.3	14.3	1.77	red
620	4.8	16.1	2.00	orange
580	5.2	17.2	2.14	yellow
530	5.7	18.9	2.34	green
470	6.4	21.3	2.64	blue
420	7.1	23.8	2.95	violet
300	10.0	33.3	4.15	near ultra-violet
200	15.0	50.0	6.20	far ultra-violet

Appendix 3. Luminescence, Fluorescence and Phosphorescence

Those who start to read the literature on luminescence are often surprised by the variation of the terminology. Luminescence is the general term of the phenomenon. We have used it in this book generally. In this way one avoids using the wrong term. The terms fluorescence and phosphorescence are especially popular among those working on carbohydrates. Fluorescence is a spin-allowed transition ($\Delta S = 0$) which is very fast, whereas phosphorescence is a spin-forbidden transition ($\Delta S = 1$) which is very slow. In compounds of the lighter elements, like carbohydrates, the spin-orbit coupling is very weak. Therefore the spin quantum numbers are good quantum numbers, i.e. there is no mixing of importance between $S = 0$ and $S = 1$ states, and the spin-selection rule is strict. In this way it is easy to distinguish between fluorescence and phosphorescence.

The terms fluorescence and phosphorescence are often used for heavier elements too. This is sometimes legitimate. For example, the $^2E \rightarrow {}^4A_2$ emission of Cr^{3+} (Sect. 3.3.4) can be called phosphorescence, and the $^4T_2 \rightarrow {}^4A_2$ emission fluorescence. The $^5D_0 - {}^7F_J$ emission of Eu^{3+} (Sect. 3.3.2.) can be called phosphorescence. It is definitely wrong to call the $^2E - {}^4A_2$ emission of Cr^{3+}, or the $^5D_0 - {}^7F_J$ emission of Eu^{3+} fluorescence, although this is frequently done. Nevertheless we prefer to call all these emissions luminescence. Actually, the $^4T_2$ L $^4A_2$ emission of Cr^{3+} is not an allowed transition, because of the parity selection rule, nor is the slow rate of the $^5D_0 - {}^7F_J$ emission of Eu^{3+} in the first place due to the spin-selection rule (the parity selection rule is of more importance in this case).

To increase the confusion, some authors use the word phosphorescence when they mean afterglow (Sect. 3.4). It will be clear that phosphorescence characterizes a specific emission transition (i.e. $\Delta S = 1$, slow), whereas afterglow relates to whatever type of emission transition after trapping of an electron or hole elsewhere.

Garlick [1] proposed, long ago, the sole use of the terminology as summarized here.

Reference
1. Garlick GFJ (1958) In: Flügge S (ed) Handbook der Physik, Springer, Berlin Heidelberg New York, Vol XXVI, p 1

Appendix 4. Plotting Emission Spectra

Experimental data on luminescent emission spectra are usually presented as relative emitted energy per constant wavelength interval, i.e. Φ_λ vs λ. Usually the maximum of this curve is considered to be the peak of the emission band.

For theoretical considerations, however, it is imperative to plot the relative emitted energy per constant energy interval, i.e. Φ_E vs E. As shown in appendix 2, the frequency (ν) or wavenumber ($\bar{\nu}$) can also be used, since they are proportional to E. The Φ_E vs E spectrum yields a value of E for Φ_E (maximum) which is not equal to the value of E which is obtained by taking the λ value for Φ_λ (maximum) and converting it to E. In the literature this is often overlooked. A simple example for illustration: if the maximum of an emission band in the Φ_λ vs λ spectrum is at 500 nm, the emission maximum is not at $\bar{\nu} = 20.000$ cm^{-1}, but at a lower $\bar{\nu}$. This can be seen as follows.

Φ_λ and Φ_E are related to each other by the equation

$$\Phi_E = \Phi_\lambda\ \lambda^2 (hc)^{-1} \tag{A 4.1}$$

Here c is the velocity of light, and h the Planck constant. Eq.(A 4.1) can be easily derived by remembering that in the Φ_λ vs λ spectrum we plot per wavelength interval $d\lambda$, i.e. we plot $\Phi_\lambda d\lambda$. In order to convert $d\lambda$ to an energy interval dE, the expression $E = hc\lambda^{-1}$ is used. By differentiation it follows that $dE = hc\lambda^{-2}d\lambda$.

As a matter of fact the maxima in the Φ_λ vs λ and Φ_E vs E spectra coincide within the experimental error in the case of sharp lines. However, in the case of broad bands they are different. Only in the Φ_E vs E plot a broad emission band is expected to show a Gaussian shape. Figure A 4.1 illustrates these effects for a simple example. The literature can be found in, for example, ref. [1]. It is self-evident, that eq. (A 4.1) should not be used for excitation spectra.

References
1. Curie D, Prener JS (1967) In: Aven M, Prener JS (eds) Physics and chemistry of II-VI compounds. North Holland, Amsterdam, chap 9
2. Blasse G, Verhaar HCG, Lammers MJJ, Wingefeld G, Hoppe R, de Maayer P (1984) J. Luminescence 29 :497.

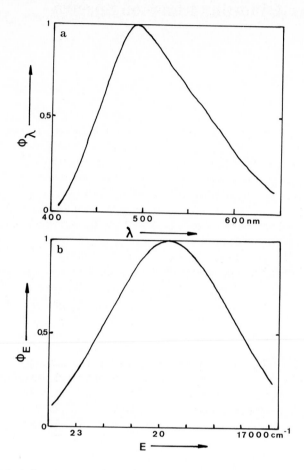

Fig. A 4.1. The emission spectrum of $Ba_2WO_3F_4$ [2] as Φ_λ vs λ (a) and as Φ_E vs E (b). The maximum in (a) is at 490 nm, and in (b) at 19 700 cm^{-1}. Note that 490 nm corresponds to $\approx$20 400 cm^{-1}

Index

The figures refer to the relevant section(s) in the book

Springer-Verlag
and the Environment

We at Springer-Verlag firmly believe that an international science publisher has a special obligation to the environment, and our corporate policies consistently reflect this conviction.

We also expect our business partners – paper mills, printers, packaging manufacturers, etc. – to commit themselves to using environmentally friendly materials and production processes.

The paper in this book is made from low- or no-chlorine pulp and is acid free, in conformance with international standards for paper permanency.

Printing: Mercedesdruck, Berlin
Binding: Buchbinderei Lüderitz & Bauer, Berlin